Evolution, Huma
and Morality

This book highlights the recent re-emergence of Edward Westermarck's work in modern approaches to morality and altruism, examining his importance as one of the founding fathers of anthropology and as a moral relativist, who identified our moral feelings with biologically evolved retributive emotions.

Questioning the extent to which current debates on the relationship between biology and morality are similar to those in which Westermarck himself was involved, the authors ask what can be learnt from his arguments and from the criticism that he encountered. Drawing on Westermarck's manuscripts and papers as well as his published work, the authors show the importance of situating debates, whether modern or classical, in their correct methodological and philosophical context.

This volume is a rigorous assessment of the ways in which morality is connected with human biological nature. It plays close attention to the development of debates in this field and will appeal to scholars of sociology, anthropology and philosophy.

Olli Lagerspetz is Senior Lecturer in Philosophy at Åbo Akademi University, Finland. He is the author of *Trust, Ethics and Human Reason* and co-author of *Edward Westermarck: Intellectual Networks, Philosophy and Social Anthropology*.

Jan Antfolk is a postdoctoral researcher in the Department of Psychology at Åbo Akademi University, Finland.

Ylva Gustafsson is a postdoctoral researcher in the Department of Philosophy at Åbo Akademi University, Finland.

Camilla Kronqvist is a postdoctoral researcher in the Department of Philosophy at Åbo Akademi University, Finland.

Philosophy and Method in the Social Sciences

Series Editor: Phil Hutchinson

Manchester Metropolitan University, UK

Engaging with the recent resurgence of interest in methodological and philosophical issues in the human and social sciences, this series provides an outlet for work that demonstrates both the intellectual import of philosophical and methodological debates within the social sciences and their direct relevance to questions of politics, ethics or policy. Philosophy and Method in the Social Sciences welcomes work from sociologists, geographers, philosophers, anthropologists, criminologists and political scientists with broad interest across academic disciplines, that scrutinises contemporary perspectives within the human and social sciences and explores their import for today's social questions.

Titles in this series

1 **Wittgenstein among the Sciences**
Wittgensteinian Investigations into the 'Scientific Method'
Rupert Read, Edited by Simon Summers
(Ashgate, 2012)

2 **Clarity and Confusion in Social Theory**
Taking Concepts Seriously
Leonidas Tsilipakos
(Ashgate, 2015)

3 **Evolution, Human Behaviour and Morality**
The legacy of Westermarck
Edited by Olli Lagerspetz, with Jan Antfolk, Ylva Gustafsson and Camilla Kronqvist

Evolution, Human Behaviour and Morality

The legacy of Westermarck

Edited by
Olli Lagerspetz, Jan Antfolk, Ylva Gustafsson and Camilla Kronqvist

Routledge
Taylor & Francis Group
LONDON AND NEW YORK

First published 2017 by Routledge

2 Park Square, Milton Park, Abingdon, Oxfordshire OX14 4RN
52 Vanderbilt Avenue, New York, NY 10017

Routledge is an imprint of the Taylor & Francis Group, an informa business

First issued in paperback 2020

British Library Cataloguing in Publication Data
A catalogue record for this book is available from the British Library

Library of Congress Cataloging-in-Publication Data
Names: Lagerspetz, Olli, 1963– editor.
Title: Evolution, human behaviour and morality : the legacy of Westermarck / Olli Lagerspetz, Jan Antfolk, Ylva Gustafsson and Camilla Kronqvist.
Description: Abingdon, Oxon ; New York, NY : Routledge, 2016.
Identifiers: LCCN 2016004025 | ISBN 9781472451743 (hardback) | ISBN 9781315581378 (ebook)
Subjects: LCSH: Westermarck, Edward, 1862–1939. | Ethics. | Evolution. | Philosophical anthropology. | Ethnology.
Classification: LCC B4715.W44 E96 2016 | DDC 171/.7—dc23
LC record available at http://lccn.loc.gov/2016004025

ISBN: 978-1-4724-5174-3 (hbk)
ISBN: 978-0-367-59615-6 (pbk)

Typeset in Times New Roman
by Apex CoVantage, LLC

Edward Westermarck (right) with Abd-es-salam El-Baqqali (left), 1909, cropped. Manuscript and Picture Unit Collections, Åbo Akademi University Library

Contents

Contributors

Jan Antfolk, PsD., Psychology, Åbo Akademi University.

Ylva Gustafsson, PhD., Philosophy, Åbo Akademi University.

Juhani Ihanus, Senior Lecturer, Adjunct Professor (*docent*), Psychology, University of Helsinki; Adjunct Professor, History of Science and Ideas, University of Oulu.

Camilla Kronqvist, PhD., Philosophy, Åbo Akademi University.

Olli Lagerspetz, Senior Lecturer, Adjunct Professor (*docent*), Philosophy, Åbo Akademi University.

Antti Lepistö, Doctoral Candidate, General History, the University of Helsinki.

Hannes Nykänen, Adjunct Professor (*docent*), Philosophy, Åbo Akademi University. Dr Nykänen works at present at the Department of Philosophy, the University of Helsinki.

Otto Pipatti, Doctoral Candidate, Sociology, the University of Helsinki.

Ullica Segerstrale, Professor of Sociology, Illinois Institute of Technology.

Kirsti Suolinna, Adjunct Professor (*docent*), Sociology, Åbo Akademi University.

Arthur P. Wolf (1932–2015), Professor of Human Biology, the Department of Anthropological Sciences, Stanford University.

Acknowledgements

This collection of essays explores the legacy and current relevance of Edward Westermarck's (1862–1939) work. It is a result of the research project *Westermarck and Beyond: Evolutionary Approaches to Morality and Their Critics*, which received three years of funding from Kone Research Foundation (2009–2011). The project itself was part of a larger research endeavour addressing evolutionary perspectives on the human sciences, also financed by the foundation. We are grateful to Kone not only for the funding, but also for the organisational effort, helping us bring together the research group in the first place and then repeatedly providing us with the opportunity to exchange ideas with members of related projects.

Most of the chapters included here have been presented and discussed regularly in the project workshops. An extended seminar on Westermarck was also organised in Åbo in 2012 with funding from The Swedish Cultural Foundation in Finland.

Apart from our interest in Westermarck, contributors to this volume are also his 'heirs' in another way. The majority of us have studied or worked at Åbo Akademi University, the Finnish academic institution where Westermarck taught in 1918–1932 and where his memory is alive in many ways. The project meetings were held in Åbo, at the Philosophy Department in Auditorium Westermarck. In particular, the bulk of Westermarck's manuscripts, correspondence and photographs are kept here. We extend our thanks to the Manuscript and Picture Unit of Åbo Akademi University Library, as well as to the Swedish-language section of the Finnish Philosophy Portal www.filosofia.fi. Finally, the readers and editors of the present series are acknowledged for advice and encouragement.

During the editing stage of this volume we received the news of the passing of one of our authors, Professor Arthur P. Wolf. Professor Wolf's anthropological work over several decades contributed significantly to the renewal of scientific interest in Westermarck. He will be fondly remembered by the worldwide community of Westermarck scholars.

Åbo, Finland, September 2015,
The Editors

1 Introduction

Westermarck and modern evolutionary approaches to morality

Olli Lagerspetz

The relation between morality and human biological nature is a research theme that holds out many promises. It is also one that must be approached with care. The present collection of essays contributes to the methodology of applying evolutionary frameworks to morality. Evolutionary psychology in its present form is a recent example of the research endeavour we will be discussing, but it belongs to an ongoing tradition.

Methodology, in brief, is about matching research questions with research methods. Thus, it concerns not only techniques of investigation in a narrow sense, but also the status of the questions posed to the material. The crucial element here is the inter-relatedness of questions, techniques and data. On the one hand, a given question suggests certain ways of looking for an answer. On the other hand, the methods at one's disposal tend to influence one's understanding of the question. In this way, settling for a method of looking is always also to settle for a particular interpretation of what one is seeking.

Relations between 'morality' and 'biology' have been a theme for several academic disciplines as well for other spheres of culture, including literature, art and religion. Philosophy, sociology, evolutionary theory, psychology and psychoanalysis have all contributed to the scholarly debate. In advancing their responses to it, the different participating disciplines have also been subtly or not so subtly reformulating the issue of 'morality and biology'. Discussing the methodology of applying evolutionary frameworks to morality thus requires addressing ways in which the crucial questions and concepts have been understood and reinterpreted – how arguments have been put forward, or sometimes just tacitly assumed, for the reformulations suggested.

Methodology in this full-fledged sense can be described as the study of how research can be conducted, understood and revisited. This means not only assessing methods in current research, but also putting past research into perspective. The editors of this volume embrace three major starting points:

1 Understanding relations between human biological evolutionary history and morality calls for *conceptual clarification* of some kind. It is at present often unclear *what* is meant by the two things that we are expected to relate to each other – 'biology' and 'morality'.

2 We believe that this kind of clarification *may* and often *must* involve a historical element. We can more easily understand the terms deployed in current controversies if we look at their development, as well as at the cultural, philosophical and political concerns that have shaped the debates over the years.
3 Finally, all contributors to this volume use the work of *Edward Westermarck* (1862–1939) as the central point of reference. Westermarck counts as one of the founding fathers of social anthropology as we know it today. He also put forward a theory of the emotional basis of morality. For him, morality was rooted in reactions of a specific kind, which in turn had a background in human biological evolution. Today Westermarck is celebrated as a precursor of evolutionary psychology. In particular, modern evolutionary psychologists have focused on his theory of incest avoidance, known as the Westermarck effect.

Methodology as self-reflexion

As to the *first* point (1): The idea of *explaining* moral behaviour is a central element in the debate on morality and evolution. But there is no agreement on exactly what should be required of a *successful* explanation of human behaviour. On this issue, we encounter the classical contrast between two conceptions regarding the status of the human sciences; between what one may call the Unity of Science view and the tradition of *Geisteswissenschaften* (see von Wright, 1971, Chapter 1). Whole libraries have been written on this conflict, and we must now confine ourselves to a brief characterisation of it. On Hempel's (1965) influential formulation of the Unity of Science view, scientific explanation consists of demonstrating that a phenomenon (the *explanandum*) is a local application of a general regularity or covering law (sometimes described as a 'law of nature'). Thus, if a specific type of human behaviour – say, altruistic behaviour – is to be explained, the evolutionary explanation might be that it is a species of the more general category of *adaptive* behaviour. The full evolutionary explanation of morality would involve showing in what ways specific types of moral behaviour are adaptations shaped by natural selection or, in some instances, by-products of such adaptations. In contrast, the tradition of *Geisteswissenschaften* describes the primary task of the humanities and the social sciences as one of 'understanding', not of 'explanation'. To put this another way, understanding takes the place of explanation because to understand fully the considerations and concerns that shape an agent's meaningful behaviour *is* to have an explanation of it. To 'explain morality' in this sense is to describe the meaning of morality in human life – finding descriptions that do justice to the concerns and considerations expressed in it.

The idea of *Geisteswissenschaften* implies a conception of the human sciences as an instance of cultural self-reflection. With regard to the task of 'explaining morality', this suggests the following view. A central part of the explanatory activity should consist of questioning received opinions about what morality is generally 'about' – not in the spirit of advocating reforms of existing moral practices, but as an attempt to see whether the moral concerns that typically engage us in our personal lives are adequately reflected in the theoretical debate. For instance, are our most important moral concerns recognisable in the kinds of general statements to which many of us are prone to assent when questions about the meaning of morality are addressed? Crucially, morality is 'about' *us*, about the meaning that we find in our own relations with others.

Conceptualisations of human behaviour in terms of evolutionary theory and a biologically determined human nature tend to bring along an entire baggage of mostly unacknowledged assumptions about the essence of morality. They typically involve the use of terms of art such as *altruism, reciprocity* and *sympathy*, to name a few central ones.[1] These words owe their first origin to quite specific theories of motivation, even though, after decades of dissemination, they have also found their way to everyday speech and popular theorising about behaviour. Nevertheless, it is safe to say that our personal understanding of the concerns and problems that shape our moral lives is both more nuanced and more chaotic than might be captured by a theory of altruism. Which leads to the question: What happens along the way when moral concerns are transposed onto an evolutionary theory of behaviour? Evolutionary explanations of morality easily run the risk of, so to speak, streamlining away the thing that they hope to explain. There are perhaps circumstances in which a rationally streamlined morality might appear a good thing, but an *explanatory* endeavour is, by definition, committed to preserving whatever is to be explained, not replacing it with an improved variety. This is, on the other hand, not to deny that philosophical and theoretical confusions are often solved by finding novel descriptions of the phenomena at issue.

Another possible response makes use of the distinction between *proximate* and *ultimate* explanations (see Antfolk, Chapter 6, this volume).[2] One might say that evolutionary theory is not supposed to explain the moral thinking of any given individual, but only the general, large patterns of human interaction and that of some of their nonhuman precursors. In other words, by applying evolutionary reasoning, scholars seek to find general laws for human behaviour – laws that may or may not apply to specific cases. For instance, Westermarck's ultimate explanation of human incest avoidance made use of the idea of a process of natural selection, but his explanation at this level did not (and as a matter of principle, could not) involve any claim to the effect that human beings as individuals *wish* to improve their fitness. Selective pressure, however, gave rise to motivational structures in individuals which in practice tend to increase their fitness. From the point of view of *Geisteswissenschaften*, however, the assimilation of individual moral motivation with causal influences at the proximate level still seems to involve conflation of causality and intentionality. For the present, we should simply note the character of the methodological disagreement. It concerns not only the preferred explanations or the preferred methods of explaining, but also the *identification* of the very phenomenon ('morality') to be explained.

By focusing on the human being as an embodied biological being, evolutionary approaches to morality have given new prominence to an idea that may serve as a corrective to traditional moral philosophy. It is the suggestion that human moral life should not be seen merely as the application of an intellectually coherent system of precepts, but rather as the outcome of natural reactions of one living being to another. This was the central motif of Westermarck's theory of the origin and development of morality. Today, neosentimentalist moral theories have capitalised on some of the insights included in Westermarck's theory. We will return to this issue in a moment.

The historical context

As to the *second* point (2): The issues just mentioned are, of course, complicated – philosophically, culturally and politically – more so than what is for the most part allowed by contemporary proponents *or* critics of evolutionary approaches to behaviour. The ramifications of these issues at the level of the general worldview can be traced by looking for cultural and other concerns that originally gave rise to the debate. This book reflects the conviction that historical analysis is essential for understanding what the discussion today is about. Without some grasp of this background of ideas, there is a distinct risk that participants of the current round of debates literally do not know what they are saying.

The attempts today to find links between evolution and morality are to some extent fuelled by the increase of technical sophistication in the applications of evolutionary theory. But the enterprise as such is not new. Evolutionism has experienced several second comings in the last 150 years or so. A persistent battle is being fought for the soul of 'real' Darwinism, and the debate constantly spills over to ideology. Because many of the arguments have remained the same over the years, we will do well to look at them in their original contexts. One thing we learn is that Darwinism is not a single ideology, but more like a toolbox, a set of resources that lend themselves to different, even contradictory, uses.

Anthropological theory developed initially, to a great extent, as a response to nineteenth-century philosophical debates concerning the 'basis', 'sources' or 'origins' of morality. For early anthropologists, the question of how morality emerged during human prehistory was at the same time a question about the *essence* of morality in the present situation. Was morality an expression of human nature, and hence 'natural'? Or should it be seen as a set of culturally imposed constraints on human nature? For instance, was sexual taboo the outcome of human natural instincts, or was it due to the repression of precisely those instincts by society? Westermarck responded to these questions by creatively applying evolutionary theory and the moral psychology of eighteenth-century British sentimentalists.

Several modern authors (such as E.O. Wilson, Frans de Waal, Leda Cosmides and John Tooby) explicitly acknowledge their indebtedness to Westermarck and claim to be developing his work. On the other hand, modern criticism of evolutionary approaches to morality is often reminiscent of the critique directed at Westermarck by his contemporaries. That includes responses by Émile Durkheim and Sigmund Freud and the mostly implicit criticism found in the work of Westermarck's student Bronisław Malinowski. At this juncture, it will be valuable to look at the roots of recent debates. Are they simply a replay of the old ones, or are there new elements? And, given the existing similarities and documented influences, to what extent are Westermarck and more recent evolutionary approaches to morality vulnerable to similar kinds of criticism?

As a preliminary answer, it seems to us that Westermarck indeed was a predecessor of evolutionary psychology. He shared many of its theoretical assumptions, including the commitment to looking for a unitary human nature, and privileging explanations in terms of instinctive emotional responses over cultural explanations. At the same time, he demonstrated a degree of philosophical awareness that could be helpful in modern attempts to address the same issues. Moreover, he was alive to the value of anthropological fieldwork.

Edward Westermarck

Finally, to the *third* point (3): a few words are needed here to explain who Westermarck was. Edward (Edvard[3]) Westermarck is usually associated with those great names of late Victorian social anthropology, Sir James Frazer and Edward Burnett Tylor (see Lévi-Strauss, [1945] 1982). All three of them were evolutionists and advocates of the approach known as the comparative method. After his graduation from the University of Helsinki (Helsingfors), Westermarck took the steamship to London in 1887, entrenched himself in the British Museum Library and started his scholarly career as an armchair anthropologist. His thoroughly researched volumes on the history of human marriage (HHM, 1891) and the origin and development of the moral ideas (ODMI I-II; 1906, 1908) gained him entry to the emerging research community of British social anthropologists. These volumes were still wholly (HHM) or overwhelmingly (ODMI) based on written sources. However, the comparison with Frazer only partly captures Westermarck's contribution to research. Unlike Frazer, he was also a fieldworker. From 1898 onwards Westermarck spent several years in Morocco and carried out field research of previously unsurpassed magnitude (summarised in RBM I-II; 1926). He was the first anthropologist to stress the importance of an extended stay in the field and of acquiring fluency in the local languages. He also passed on these insights to his students. Westermarck's anthropological life work thus has both traditional and forward-looking aspects. On the one hand, he used the comparative method to create syntheses from a large mass of secondary sources in the style of his nineteenth-century predecessors. On the other hand, he conducted meticulous research of his own with local informants.

Through most of his adult life, Westermarck held teaching positions in both Finland and Great Britain at once. He worked as a professor of philosophy at the University of Helsinki (1906–1918) and Åbo Akademi University (1918–1932) and as a teacher of sociology at the London School of Economics and Political Science (lecturer 1904, professor 1907–1930). At the London School of Economics he was strategically placed at a time when social anthropology was a novel discipline, soon to expand vigorously in Britain and in the Commonwealth. Many of his students became prominent social scientists (including Bronisław Malinowski, Raymond Firth, E.E. Evans-Pritchard, Talcott Parsons, Rafael Karsten, Gunnar Landtman, Uno Harva and Hilma Granqvist). However, his influence was eclipsed by that of his star pupil, Malinowski. Modern treatments of the history of social anthropology tend to downplay Westermarck's role in shaping the discipline. In Westermarck's declining years, the younger generation already considered his evolutionism to be a thing of the past. The evolutionists were above all criticised for their neglect of social context. 'New anthropology' no longer concerned itself with the past origins of social institutions, but with their function in their present environment.

In Westermarck's youth, the social sciences had not yet properly gained their independence from philosophy. His comparative work on the development of morality (ODMI) starts with a theory of the moral concepts; the subsequent chapters are designed as detailed illustrations of the theory. Westermarck's later work *Ethical Relativity* (ER, 1932a) is a restatement of his position in the earlier book, now with more focus on philosophical argument. Westermarck argued that ethics cannot be a normative discipline. Normative ethical systems that claim rational grounds, such as Utilitarianism and Kantianism, in fact, reach their conclusions on the basis of feeling. Scientific ethics must be reconstituted as the study of the emotions that give rise to moral judgements. Westermarck belongs to the few authors to have presented an elaborate philosophical defence of relativism. Most philosophers would rather see relativism as the obvious thing to avoid. In contrast, in social anthropology, some form of relativism is often believed to be more or less implicit in the discipline as such (see Österman, 2014, pp. 97–103). It is possible that Westermarck contributed to this situation. An educator of a generation of British social anthropologists, he naturally passed on his philosophical and methodological views to them. On the other hand, Westermarck's support of relativism no doubt partly reflected his fieldwork experience. In his introduction to the French edition of ODMI (1928–29) Westermarck claimed that his numerous anthropological examples were not simply illustrations of his theoretical position, but the empirical basis from which the latter was derived (see also Lagerborg, 1951, p. 136). The present volume takes up the question how to understand Westermarck's version of relativism. In any case, his philosophical and social anthropological concerns were closely intertwined: his dismissal of rationalist ethics motivates empirical research, whereas his empirical examples highlight the role of emotions in the formation of moral concepts.

Morality, according to Westermarck, is relative because it is relative to feeling. Morality emerged from what he termed retributive emotions – instinctive reactions with a background in biological evolution. The basic feature of moral life consists of our tendency to pay back in kind: we wish to inflict pain in response to pain and to give pleasure in response to pleasure. These impulses turn into *moral* responses when they incorporate claims to impartiality, disinterestedness and universal validity. This means that we think of our responses as embodying normative expectations: we think of them as the *right* reactions.

In one key respect, Westermarck's theory of the evolution of morality is strikingly different from the theories now usually suggested in the debate on morality and evolution. The problem most frequently addressed today is how the emergence of (selective) altruism is possible – whereas egoism of some kind is seen as the natural default position *not* in need of explanation. Westermarck, however, does not frame his theory in terms of this contrast, but in terms of hostile and benevolent responses, both of which he sees as *equally* natural. The central problem for him is how it came about that animal instinctive responses later, in the course of cultural evolution, acquired the character of normativity. Normativity, not altruism per se, is what distinguishes moral judgements from mere emotional responses.

The last few years have seen a partial 'resurrection' of Westermarck's work by modern researchers. So far, however, resurrection attempts have overwhelmingly focused on a limited motif in his work, namely, his theory of incest avoidance. Although the contributors to the present volume by no means agree in their views on evolutionary psychology, we agree on two things: first, Westermarck had more than this one string to his bow; second, the current discussion on the evolutionary background of morality would profit from taking up a historical perspective and studying its own predecessors.

The four parts of the present volume

Edward Westermarck made contributions to what we would today treat as several academic disciplines: to social anthropology, psychology and philosophy. Moreover, he was a predecessor of evolutionary psychology – itself a field of research difficult to classify, because its present form incorporates elements from many areas, including individual psychology, rational choice theory and population genetics. The subsequent chapters of this volume will each focus on some specific aspect of Westermarck's oeuvre. The book is organised in four thematic parts, each starting with an introductory background discussion.

Part I, *Westermarck and the Emergence of Twentieth-Century Social Anthropology*, relates Westermarck's work on social anthropology to the main theoretical and methodological challenges that he encountered in this area. As the introductory chapter by Olli Lagerspetz (Chapter 2) sums up, challenges came mainly from three quarters: (1) the advocates of the geographically limited monograph as the format for anthropological research; (2) functionalism; and (3) psychoanalysis. Westermarck's answer to the first challenge was to go along with it and contribute substantially to it, while still retaining the ideal of the comparative synthesis as the essential long-term goal for anthropological research. Westermarck's responses to the two other challenges were characterised by polemic and rejection.

The essay by Juhani Ihanus (Chapter 3) describes Westermarck's relations and disagreements with the psychoanalytic movement, including an outline of Malinowski's role in the debate. The main conflict between Westermarck and psychoanalysis concerned the Oedipus complex and the incest taboo. Although Westermarck did not actively question the therapeutic usefulness of psychoanalysis, he believed its applications to social anthropology were merely speculative. The chapter by Olli Lagerspetz and Kirsti Suolinna (Chapter 4) takes up Westermarck's response to Durkheim's functionalist theory of religion on the basis of his Moroccan field results. The authors argue that Westermarck's work on Moroccan folk religion includes an analysis of holiness, or *baraka*, which implicitly identifies it as an aspect of politically contested power relations. Thus his fieldwork pointed beyond both Durkheim's and Frazer's construals of religious belief. The following chapter by Lagerspetz (Chapter 5) is a critique of Westermarck's cross-cultural comparisons in light of the ideas of social understanding later developed by Peter Winch. A question that emerges is to what extent similar criticism is applicable to modern evolutionary approaches to human behaviour.

Part II – *Westermarck as a Precursor of Evolutionary Psychology* – addresses applications of Westermarck's theory of incest avoidance in modern research. Chapter 6 by Jan Antfolk introduces the theme. Westermarck's most debated contribution to research today is known as the Westermarck effect, relating to incest aversion and incest prohibition. Westermarck claimed that natural selection had shaped human sexuality so as not to include as sexual partners individuals with whom the agent has coresided in his or her early youth. As such individuals tend to be close kin, this aversion manifests itself as aversion against sexual relations with close kin. Westermarck's explanation privileges neither 'nature' nor 'nurture', but highlights their interplay in human development.

Antfolk and Arthur P. Wolf (Chapter 7) point out that the debate on Westermarck's theory of incest aversion has been blurred in focus, mainly because it is mistakenly seen as one single hypothesis. The authors argue that 'the Westermarck hypothesis' consists of four different, logically distinct claims on separate explanatory levels. The chapter by Ullica Segerstrale (Chapter 8) follows Edward O. Wilson's handling of the Westermarck effect in his books from 1975 to 2012. Over the years Wilson is moving from using the Westermarck effect as a straightforward example of the workings of natural selection on behaviour, toward a consideration of more complicated connections between proximate and ultimate causes. Segerstrale also addresses the puzzling fact that Wilson, although aware of Westermarck's argument and using it for his own purposes, chose for a number of years to downplay Westermarck's role as its inventor.

The Westermarck effect, then, has been very influential for sociobiological theorising: both for the idea of gene/culture coevolution (as an example and thinking tool) and as some kind of glue or connecting link for the unification of knowledge. This idea, therefore, has taken on a methodological and philosophical power far beyond what its original creator could have dreamt of.

Part III – *The Emotional Origins of Morality* – places Westermarck's theory of morality into the context of the sentimentalist and emotivist movements that occurred in philosophy both before and after his lifetime. The introductory essay (Chapter 9) is written by Camilla Kronqvist and Otto Pipatti. The central philosophical questions addressed by Westermarck belong to meta-ethics, that is, the study of the logical status of moral judgements. However, Westermarck believed that the task of scientific ethics must be descriptive and include empirical study of existing forms of moral consciousness. These two aspects of his thinking reflect a tension between philosophical or conceptual methods of clarifying moral language and, on the other hand, more empirically minded approaches to explaining moral psychology. This tension is not fully resolved in Westermarck's writings, and it continues to haunt more contemporary descriptive attempts at explaining morality through their emotional basis.

In Chapter 10, Pipatti traces the development of Westermarck's theory of morality during his early career. Westermarck originally followed Darwin's views concerning the evolution of social instincts. His closer acquaintance with Adam Smith's *Theory of Moral Sentiments* caused him to modify his views, giving pride of place to the notion of sympathy. Sympathy acquired a moral character due to the social context described as the General Will. Finally, Westermarck combined his previous accounts of moral approval and disapproval on the one hand, and resentment and gratitude on the other, into a single unified theory of moral emotions as retributive emotions. Chapter 11 by Kronqvist addresses the debate between Westermarck and G.E. Moore on the logical status of moral judgements. Both thinkers subscribed to the Humean distinction between 'is' and 'ought', but their conclusions differed. The crucial question concerned moral disagreement. Moore argued that the existence of moral disagreement was proof of the existence of moral facts. Westermarck claimed, however, that human moral thinking displays a considerable unity beneath surface disagreements. On the other hand, he held that the impossibility of reconciling real moral differences reveals their emotional character. Where Westermarck's discussion does appear to have a considerable advantage over Moore's treatment of moral judgements, however, is in his realisation that moral emotions are *forms of reaction* we direct at each other, and thus that they are characteristically set in an interpersonal context.

Westermarck may be seen as a precursor of current neosentimentalist theories of morality. In Chapter 12, Kronqvist considers Westermarck in relation to the arguments for moral relativism recently raised by Jesse Prinz. Prinz hopes to develop Westermarck's sentimentalist and relativist approach, but he advances a sentimentalism that at several important junctures differs from that of Westermarck's work. Prinz is critical of evolutionary psychology because of its tendency to underplay the significance of culture in the construction of morality. Yet he shares its commitment to a reductive form of naturalism, involving the assumption that the *facts* of human moral behaviour may be described in a language that is independent of normative moral concepts. Kronqvist asks, however, whether such attempts to empirically investigate morality succeed in exonerating the normativity inherent in all moral inquiry.

Part IV – *Evolutionary Psychology and Morality: Critical Perspectives* – focuses on ways in which evolutionary studies of morality may in fact result in *changes* rather than *elucidations* of key moral concepts. In her introduction (Chapter 13), Ylva Gustafsson outlines the development. Scientific studies of moral behaviour have tended to replace everyday moral vocabulary, signifying blame or approval, with one that implies focus on merely formal qualities of behaviour (such as 'altruism' or 'reciprocity' as descriptions of animal behaviour patterns). Familiar words that generally imply blame (such as 'cheating' or 'egoism') are employed as terms of art, taken to be value neutral. That notwithstanding, evolutionary arguments are at times fed back to general debates on human nature and assumed to have moral and political implications.

Chapter 14 by Gustafsson scrutinises the use of the concept of altruism in current debates on the evolution of morality. A main question there has been how to explain the possibility of altruism. The contributors draw on two main philosophical traditions: that of sentimentalism, emphasising the role of empathy (now developed by Frans de Waal) and the rationalist tradition originating in the writings of Thomas Hobbes (now prominent in the evolutionary psychology of Leda Cosmides and John Tooby). Westermarck has been an important influence on both research approaches, even though his views stand closer to the former than the latter. However, Gustafsson argues that the concept of altruism may simply not be an adequate way to capture the varieties of other-regarding relations characteristic of the moral life. In Chapter 15, Hannes Nykänen emphasises the Kantian features evident in both Westermarck's work and current naturalist and evolutionary accounts of the role of feelings in morality. Kant describes feelings as the 'natural' part of human psychology and, as such, as both devoid of understanding and as fundamentally self-regarding. For Kant, this creates a difficulty about accounting for the feeling of respect for the moral law, an emotion that nevertheless has a central role in his description of moral motivation. It also fails to account for sympathy, which is, by definition, other-regarding. Nykänen argues that, although Westermarck rejected Kant's project of giving morality a foundation in practical reason, he retained Kant's view on what morality *is*: a system of thought mainly characterised by impartiality and resistance to subjectivity. Nykänen also argues that modern evolutionary approaches to morality have inherited these confusions, resulting in the impossibility of accounting for truly other-regarding emotions.

A question that looms large on evolutionary studies of human behaviour is whether such studies have political implications. Westermarck was personally a man of liberal, mildly progressive leanings. The essay by Antti Lepistö (Chapter 16) is an overview of the use of Westermarck's work by North American conservative thinkers. In particular, his theories concerning parental emotions, mating patterns and the incest taboo have been appropriated for political purposes by neoconservatives looking for Darwinian backing for their ideas on family life and the role of government. Westermarck's works are, and have always been, open to various types of political use, at times different from what he probably could have imagined. But, as Lepistö points out, such openness to interpretation and exploitation of course holds true of scientific writing in general.

The contributors to this volume represent a variety of disciplines – anthropology, history, philosophy, psychology and sociology – which itself is a reflection both of the broad range of themes that Westermarck addressed and of the compartmentalisation of present-day academia. A unifying thread of the volume is the attention to research method. Methodological self-reflection also involves awareness of historical context. Westermarck was a *precursor* of evolutionary psychology. However, although he was certainly (among other things) a behavioural scientist who wished to apply Darwin's theory of evolution, he was obviously not an evolutionary psychologist in the modern sense. Moreover, he was a moral philosopher, and his views on the evolution of morality cannot be understood apart from his analysis of the moral concepts. He put forward a conception of the relation between emotion and morality that in his own time was a minority view, but which is both echoed and paralleled in philosophy today.[4]

Notes

1 On 'altruism' as a technical biological term, see e.g., Ruse and Wilson (1986), p. 179.

2 In the debate, reference is often made to Tinbergen (1963) who introduced this distinction to the study of animal behaviour. On the other hand, as Antfolk points out (this volume), a large number of philosophers from Aristotle onwards have embraced somewhat similar distinctions between different concepts of causality.

3 The official spelling of Westermarck's first name was *Edvard*. When he wrote English, he consistently preferred the English spelling (Edward) while retaining the original *Edvard* when he used Swedish, his first language.

4 In writing this chapter I have profited greatly from suggestions and comments by my co-editors – Jan Antfolk, Ylva Gustafsson and, especially, Camilla Kronqvist.

Part I

Westermarck and the emergence of twentieth-century social anthropology

Sub-editor: Olli Lagerspetz

2 Westermarck and the emergence of twentieth-century social anthropology

Three main challenges

Olli Lagerspetz

Edward Westermarck's original claim to fame rested on his large comparative studies of sexuality and moral beliefs. In those works he developed the view that morality in general was based on emotion, ultimately derived from instinct. With regard to religion, Westermarck's approach was largely the same: he believed that religion was based on instinctive feelings of awe and wonder. His overall ambition was to create a vision of human nature underlying all cultures from 'savage' to 'civilised'.

Westermarck saw himself as faithfully continuing the evolutionist tradition of his Victorian immediate predecessors. During the first decades of the twentieth century, however, that tradition was increasingly being questioned. Challenges came mainly from three quarters: (1) the advocates of the geographically limited monograph as the format for anthropological research; (2) functionalism; and (3) psychoanalysis. Westermarck's answer to the first challenge was to go along with it and contribute substantially to it, while still retaining the ideal of the comparative synthesis as the long-term goal. Westermarck's responses to the two other challenges were characterised by polemic and rejection.

The next three chapters focus on challenges to Westermarck's anthropological method. The contribution by Ihanus describes Westermarck's relations and conflicts with the psychoanalytic movement, including an outline of Bronisław Malinowski's role in the debate. The chapter by Lagerspetz and Suolinna takes up Westermarck's response to Durkheim's functionalist theory of religion on the basis of his Moroccan field results. The chapter by Lagerspetz is a critique of Westermarck's cross-cultural comparisons in light of the ideas of social understanding later developed by Peter Winch (1926–1997). The present introductory essay briefly presents the early challenges to Westermarck's method.

From large-scale syntheses to geographically limited monographs

The year 1898 saw the beginning of Westermarck's lifelong fieldwork efforts in Morocco. As a significant coincidence, it was also the year of the Cambridge expedition to Torres Strait. The Torres Strait expedition under the leadership of Alfred Cort Haddon (1855–1940) was a systematically organised interdisciplinary endeavour that became a future milestone in the development of social anthropology. The outcome was a model for others and an argument for an expansion of the field (Herle & Rouse, 1998; Rouse, 1998). Several other members, notably W.H.R. Rivers and C.G. Seligman, were also later to contribute to the development of social anthropology in Britain. Haddon and Seligman would work as Westermarck's colleagues in London.

There was a difference between Haddon's and Westermarck's initial motives, however. Westermarck appears at first to have been unsure of what his primary task was to be in Morocco. When he travelled from Tangier to Tetuan in the beginning of his fieldwork period, he brought with him seven mules to carry 'almost [his] entire philosophical library and all [his] papers' over the mountains (Westermarck to Lagerborg, 27 January 1899, ÅAB: EW). Westermarck assumed that popular beliefs and superstitions would corroborate his philosophical work as examples of how moral ideas arise from the emotions. However, as his knowledge of popular religion in Morocco grew deeper, he came to think of it as a subject in its own right.

At Torres Strait, Haddon had in mind a kind of musealisation, later to be known as salvage ethnography. He wanted a complete documentation of the local cultures, because traditional societies in the area seemed to be heading for rapid extinction. Expedition members not only conducted interviews and collected artefacts, but they also carried out psychological tests, created film documents with the newly invented cinematograph and organised re-enactments of native ceremonies earlier suppressed by the local Christian congregation. In a letter to Westermarck in 1908, Haddon summed up his view on the tasks of anthropology:

> The time has passed when students were satisfied with general accounts of native races made by the passing traveller or the untrained and frequently unsympathetic observer. Our watchword must now be 'the intensive study of limited areas'. We require to know all the conditions of existence of a given people, – how the environment affects them; how they react on it. But above all we need an accurate and exhaustive study of the psychology, sociology and religion of the people studied.
>
> (Haddon to Westermarck, 20 July 1908, ÅAB: EW)

The document is a copy of Haddon's recommendation in support of funding for Westermarck's two Finnish students, Rafael Karsten and Gunnar Landtman. By this time Westermarck had spent several years in Morocco and was fully appreciative of the value of intensive fieldwork. He encouraged his students to venture into field studies, resulting in the kinds of monographs that Haddon wanted. With his work on Morocco Westermarck also managed for his own part the transition from comparative studies to field anthropology. In his memoirs Westermarck conceded that he did 'see the weak points of the comparative method without the need of reminders from others' (MML, p. 300). Large-scale syntheses require that the author should master an enormous amount of material. Their main weakness, as he saw it, was that they tended to detach cultural phenomena 'from the organic whole' of the local culture.

> [S]ocial phenomena are not isolated facts, but are largely influenced by local conditions, by the physical environment, by the circumstances in which the people are living, by their habits and mental characteristics. And all these factors can of course be taken into account much more easily when the investigation is restricted to a particular group of men than when it comprises things that are common to widely separated peoples or to mankind in general.
>
> (MML, pp. 298–9)

Nevertheless, Westermarck still saw the comparative synthesis as the desired end result. He justified himself by suggesting that different methods should complement each other. Fieldworkers should produce monographs that could later be brought together as elements of a synthesis (Westermarck, 1936, p. 248; MML, p. 303).

In his fieldwork, Westermarck described the customs among different tribal communities in Morocco. His material concerned popular beliefs about supernatural events. As folklore, it sometimes related events in the distant past, and sometimes more recent cases, but always as retold by contemporary informants. There is no need to question the solidity of Westermarck's Moroccan studies even if more modern work is obviously now available.[1] Yet in retrospect probably their most lasting benefits lay in the development of fieldwork methods and in the fact that he passed on his experience to students. Westermarck relied on local experts for information and instruction. Only members of the relevant tribe would be relied upon – not outsiders and never Europeans. One of the first anthropologists to stress the need of acquiring fluency in the local languages, he was able to work without an interpreter.

Both as a teacher and through his own example, Westermarck contributed to defining the conventional career path expected of a student of social anthropology. The would-be anthropologist should follow a route via formal university studies in combination with a *rite de passage* consisting of an extended period of field research. With this standardisation of education practices, social anthropology had fully established itself as a discipline.

Functionalism vs. comparative method

The first influential criticism of Westermarck's comparative method was voiced by Durkheim in his review of *The Origin and Development of the Moral Ideas* in 1907. Even afterwards, Durkheim and the functionalist sociology launched by him remained a source of constant irritation for Westermarck. Durkheim wrote,

> Every time he [Westermarck] makes a claim, he cites examples from most disparate societies for proof. He is above all busying himself with amassing facts [. . .]. Not even critique of the documents thus employed is possible: receiving

> them from so many different hands, one must simply accept them as offered. [...] For him, there are no qualitatively different types of morality corresponding to similarly different social environments; at bottom he evidently believes in a morality which is one and only, inscribed in congenital human nature, and of which the moralities known to us from ethnography and history are mere progressive approximations.
>
> (Durkheim, 1907, pp. 385, 394, translation by the present author)

Durkheim's dismissal of Westermarck's ethnographic cherry picking is simultaneously a statement of his own research programme: he believed one should explain social institutions by looking for their contributions to the life of the community as a whole:

> [i]n order to discover the causes of a given moral rule and explain its historical variations one must necessarily put these variations in relation to the social environments where the rule has arisen and been subject to changes. To divorce them is to cut the moral rule off from its living source; it is to make it impossible to understand them.
>
> (Durkheim, 1907, p. 384)

The meaning of an institution or a custom was, according to Durkheim, mainly to be seen in its contribution in maintaining the particular kind of division of labour, or form of solidarity, characteristic of the society in question.

Malinowski, Westermarck's student and friend, always maintained a good relationship with his former teacher and seldom criticised him specifically (see Ihanus, this volume; Lagerspetz & Suolinna, 2014, pp. 64–5). He was quite critical in his review of Durkheim's anthropological work *Les formes élémentaires de la vie réligieuse* (Malinowski, 1913; cf. Durkheim, [1912] 1968). This did not, however, prevent him from developing the functionalist approach on his own. In 1912 Malinowski contributed to Westermarck's fiftieth anniversary *Festschrift*. His chapter dealt with the economic aspects of the *intichiuma* ceremony (Lagerspetz & Suolinna, 2014, p. 63; Malinowski, 1912). Australian totemism, of which this ritual was the most famous example, had become a focal point of international anthropological attention. Malinowski, however, applied a new perspective to it. He noted that *intichiuma* required serious preparation with intensive work, food rationing and division of labour – arrangements otherwise absent within the community. This form of magic had, in other words, a direct and real relevance to the tribal economy, completely independent of its content or validity as a worldview. He further speculated that similar ceremonies in more favourable physical environments might have been the initial impulse for the development of societies characterised by permanent division of labour. The essay, published before Malinowski set out to do his own field research, already showed features typical of his mature work. Malinowski did not speculate about the origins and spread of the ceremony. He proposed a kind of functional analysis (without actually using the word) – highlighting an aspect of the magical ceremony which did not directly reflect its content of ideas but which, as a kind of side effect, contributed to the welfare of the community. Furthermore, he spelled out the significance of *intichiuma* in terms of its impact on social organisation. These are, of course, typical themes of functionalist anthropology. Malinowski made somewhat similar observations ten years later in his book *Argonauts of the Western Pacific* (1922a), the work on which much of his subsequent fame rested.

At around this time Malinowski is likely to have become acquainted with (Sir) Alan Gardiner, the Egyptologist and linguist (Young, 2004, pp. 551, 598).[2] Gardiner was a friend of Westermarck's and a relative through marriage. Gardiner's thoughts on language influenced Malinowski's way of thinking as well as his fieldwork reporting style. The fieldworker should, when possible, reproduce the words of individual informants directly and not, as was the earlier practice, digest their responses into general descriptions. Linguistic scholarship should focus not on semantics, but on utterances and their pragmatic context – an idea reminiscent of Ludwig Wittgenstein's thinking a few years later (Young, 2004, p. 599). Here once more we see a development in anthropology towards greater attention to local context instead of cross-cultural similarities.

Malinowski never allowed the tension between his views and those of Westermarck to reach a breaking point. Their continued close relationship is attested in their correspondence, their joint teaching in the 1920s and the favourable reviews they wrote of each other's works (Malinowski, 1922b, 1927a; Westermarck, 1913). As to Durkheim and the Durkheim school proper, Westermarck always believed they were generalising from a scant empirical basis (see Lagerspetz and Suolinna, this volume). Initially Westermarck was inclined to dismiss functionalism as just a parochial French invention. In the 1920s, when also British anthropologists started taking functionalism seriously, he had to revise his stance slightly. In his presidential address on the occasion of the visit of Marcel Mauss to the University of London in 1928, he was prepared at least to pay lip service to the view that the 'English' and 'French' schools of sociology (properly put, evolutionist and functionalist anthropology) had something to learn from each other (Lagerborg, 1953, pp. 19–20; Lagerspetz & Suolinna, 2014, p. 92).

Anthropology, philosophy and psychoanalysis

During Westermarck's early career, social anthropology had not yet been clearly demarcated from the neighbouring disciplines of philosophy and psychology. For early anthropologists, the quest for 'origins' was a response to philosophical and existential questions about the *nature* of 'man'. It was to explain the meaning of morality also at the present moment, contributing to a lively philosophical debate at the time (e.g., Nietzsche, [1887] 1981; Schopenhauer, [1841] 1919). For instance, the origin of exogamy in Australia is a matter that might seem peripheral on its own. For many researchers with partly opposing views (Westermarck, Durkheim, Freud and Malinowski), it nevertheless became a joint vehicle on their road to fame. They managed to connect the theme to extra-scientific or only partly scientific issues, including questions such as: Is there anything that is absolutely right or wrong? Is religion a foundation of society, or will it be a thing of the past in an enlightened civilisation? Is man's relation to culture characterised by harmony or subjection? Is sexuality a constructive or destructive force? Their central source of information was the research on Central Australian totemism, especially among the Arunta, by W.B. Spencer and F.J. Gillen (1899). For decades after its publication, Spencer and Gillen's work was used for a plethora of purposes. It had the rare distinction of being an authoritative description of a group of people who, by Western academic consensus, represented the shared distant past of man. Totemism, in Durkheim's words, supposedly represented 'the most primitive and simple religion which it is possible to find' (Durkheim, [1912] 1968, p. 95). Similarly, in his *Totem and Taboo* Sigmund Freud told his readers that Australian totemism embodied 'a well-preserved, early stage of our own development' (Freud, [1912–13] 1946, p. 3).

According to Freud, the explanation of taboo – of which exogamy in Australia was the crucial example – may 'in the end throw light upon the dark origin of our own "categorical imperative"' (Freud, [1912–13] 1946, p. 32); he thus connected taboo with an interpretation of Kant's views on the essence of morality (Kant, [1785] 2002). Like taboo, modern morality still 'tends to act compulsively and rejects all conscious motivations' (Freud, [1912–13] 1946, p. 10). However, according to Freud, the concept of taboo remained unexplained only as long as merely its *present* forms were considered. He developed the idea that the beginnings of human culture lay in the introduction of taboo, the earliest form of which was the prohibition of incestuous sexual relations (see Ihanus, this volume).

Westermarck regarded Freud's theory of the incest taboo as completely speculative. In a letter to his closest Finnish colleague and former student, the philosopher Rolf Lagerborg (1874–1959) he made fun of its 'monstrosities' or 'oddities'[3] (Westermarck to Lagerborg, 17 June 1929, ÅAB: EW). Nevertheless it was obviously an emotional strain on him to see both Lagerborg and Malinowski show interest in psychoanalysis. Westermarck's book *Three Essays on Sex and Marriage* (1934b) included his late defence of his own theory of incest against Freud's (see Ihanus, this volume). He never made any attempt at the kind of half-hearted *détente* he achieved with functionalism.

Westermarck and the development of anthropology

After the First World War, evolutionism was increasingly being seen as outdated. This was partly a generational issue: Westermarck and his peers continued to work within the old paradigm and had no difficulty in finding respectable publishers, but the young were looking elsewhere for inspiration. Taking a bird's-eye view, it appears that the old and new paradigms also differed on the larger issue of the relation between human nature and society. Functionalists and Freudians inclined towards a view of society as an independent force – for Freudians, a repressive force – exercising control over the minds of individuals, whereas the evolutionists tended to think of social norms as more or less direct expressions of individual psychology. On the other hand, they also agreed on a number of issues, including a general belief in moral relativism and the assumption that contemporary human societies could be arranged on an evolutionary scale from savage to civilised. That was the rationale for their continued belief that research on 'savage' customs and beliefs would throw light on the origins and consequently, as they saw it, on the *essence* of human culture.[4]

Notes

1 For a modern description of the role of Moroccan saints, see Rosen (2002, pp. 75–82); for a brief description of the development of anthropological scholarship on the Maghreb after Westermarck, see Rosen (2002, pp. 41–43). Modern assessments of Westermarck's work on Morocco are included in Bourquia and Al Harras (1993).

2 On the probable date of Gardiner and Malinowski's first meeting, see Young (2004, p. 598) and Lagerspetz and Suolinna (2014, pp. 61 n314, 122). After the war, Malinowski 'quite often came to stay' with the Gardiners (Margaret Gardiner to Knut Pipping, 18 March 1976, ÅAB: KP, Box 4, Folder 2).

3 In Swedish: *vidunderligheter*.

4 Kirsti Suolinna is gratefully acknowledged for expert advice, suggestions and comments, as well as for her generous help with identifying relevant sources. We have also earlier dealt with related themes in our joint work, especially Lagerspetz and Suolinna (2014), from which some passages of this essay are adapted.

3 Westermarck, Malinowski and the 'wild things'

At the interface between anthropology, sexology and psychoanalysis

Juhani Ihanus

In his memoirs, Edward Westermarck mentions that he began to study the history of marriage almost accidentally and without any personal interest in the topic. He was early on interested in the origin of sexual modesty,[1] but later expanded his research to include the broader field of sexual relations and practices as a whole. The original topic became only a part of the wider theme. Marriage and sexuality (understood in a broad sense) overlapped in Westermarck's research throughout his career. He also never stopped thinking about the problems of sexual research and the future of marriage and focusing on the interconnectedness of biological, psychological and sociological research.

Bronislaw Malinowski was, in the context of British social anthropology, widely read in Continental scientific scholarship, and he recognised Westermarck along with Durkheim and Steinmetz as founders of sociology as an empirical science. Malinowski's anthropological and sociological training was completed under Westermarck's direction, and his interests in sexuality, marriage and kinship were informed by Westermarck's model of theory, fieldwork, philosophical detachment and captivating literary style. Besides that, Alexander F. Shand's theory of sentiments, emotions and impulses, Freudian psychoanalysis and finally behaviourism were Malinowski's great influences.

Sexology, psychoanalysis and anthropology

The initial stages of Westermarck's research career coincided with the pioneering stages of research on sexuality, or 'sexology', and psychoanalysis. Westermarck's work was linked to both areas of research, but he took a more positive view of sexology than of psychoanalysis. He also related the achievements of anthropological research to the framework of sexology and psychoanalysis. The present chapter deals mainly with Westermarck's and partly with Malinowski's connections to sexuality research, psychoanalysis and the Oedipus complex debate at the interface between anthropology, sexology and psychoanalysis.[2] What I call here the 'wild things' include sex, instincts, emotions, complexes, conflicts, incest, promiscuity, 'free love' and their manifestations and repressions in 'savage society', as researched and interpreted by two eminent immigrant social anthropologists on British soil.

When Malinowski first landed on New Guinea in September 1914 to work among the inhabitants of Mailu, Westermarck had already visited Morocco a dozen times. As Firth (1981, p. 123) has noted,

> Malinowski had the example of a man he respected, a theoretical sociologist as well as a vernacular-using fieldworker. In a sense, then, he may be said to have been following one thread of the European Orientalist tradition in his immersion in the field.

To illustrate the tone of Malinowski's gratitude toward his teacher, mentor, colleague and friend, a quote from Malinowski's letter is telling:

> You know how deeply grateful I feel for all you have done in the past; for your initial help, when I was quite alone & unbefriended in London; for your advice & assistance in bringing out my book [*The Family among the Australian Aborigines*, 1913], most of all for your personal friendship with which you have honoured me. I am very conservative in my friendships & I value yours very highly for many reasons, not the least because our scientific & general ideas & aims are very much in harmony. The war has spoilt & distroyed [sic] many things & many personal relationships. But as we two have as similar ideas & feelings about almost everything, why should we be brought one inch apart?
>
> (Malinowski to Westermarck, Tenerife, 18 July 1921, ÅAB: EW, Box 22, number 3454)

There are many other mutually thankful, and even laudatory, statements in private letters and in publications by both of these men.

As anthropological fieldworkers and scholars, Westermarck and Malinowski continued the tradition of nineteenth-century classics that observed, depicted, discussed and assessed sexual beliefs, customs and practices, primitive kinship institutions and relations between parents and children and between other kinsfolk. For example, the controversy about primitive promiscuity and group marriage can mainly be related to the question of the modes of sexual conduct and the ways of relating sex, parenthood and children. Psychoanalysts referred in this regard to the topic of the 'primal scene' (cf. Fortes, [1957] 1968, p. 167).

Psychoanalytic anthropology and the dispute over the Oedipus complex

The formation of psychoanalytic anthropology

Freud's *Totem und Tabu* (1912–1913)[3] opened the way for the application of psychoanalysis in the examination of anthropological, mythological and religious issues. Abraham, Riklin, Jones, Jung, Silberer, Rank and Reik were among the first of Freud's students to approach anthropological and mythological issues and questions on religious practices from a psychoanalytic angle. The Hungarian Géza Róheim was the first anthropologist to adopt the psychoanalytic method. In the late 1910s, Róheim developed psychoanalytic anthropology[4]; he was also the first anthropologist who had gone through psychoanalysis (by Ferenczi during the First World War and, later, by Vilma Kovács) to do fieldwork (with his wife, Ilonka, from the end of 1928 to the spring of 1931).[5]

Early British anthropologists avoided psychoanalytic theory, with the exception of W.H.R. Rivers, C.G. Seligman and, to some extent, B.Z. Seligman. Rivers, who had been educated in psychology and psychiatry, became interested in Freud's writings during the First World War. In 1916 he paid homage to the psychoanalytic movement 'which, in spite of all its faults, I am inclined to regard as one of the most important which has ever taken place in the history of the science' (Rivers, 1916, p. 12). Freud's theory had revealed the importance of the unconscious and 'the misleading character of the intellectual motives by which the actions of mankind are usually explained' (Rivers, 1916, p. 13). Such words in praise of psychoanalysis were rare among British anthropologists of the 1910s. In 1918 C.G. Seligman urged Malinowski to study psychoanalytic theory and to apply it in fieldwork. He encouraged Malinowski to, for example, collect dreams from the informants on the Trobriand Islands (see Stocking, 1986, p. 31). Nevertheless, Malinowski's closer acquaintance with psychoanalytic writings was postponed until the early 1920s.

Westermarck mentions Freud in the revised and enlarged fifth edition of *The History of Human Marriage* (1921a), making a disparaging remark about how both Frazer and Freud criticised his theory of exogamy. Westermarck believed that the same arguments he had used to counter Frazer's opinions on exogamy would suffice to overrule Freud's views. In a footnote, Westermarck makes plain his reservations about psychoanalysis, an attitude that was to remain largely unchanged:

> That the results of the so-called psycho-analysis are destructive to my theory is a supposition for which I must see some evidence before I can take it seriously. In Dr. Freud's terminology, 'sexuelle Regungen' may imply mental states very different from the desire for sexual intercourse; and the study of neurotic persons can hardly be regarded as a safe guide to the proper understanding of the normal manifestations of the sexual instinct.
>
> (HHM II, 1921, p. 204; see also HHM II, 1921, p. 192 (footnote 1); SHM, pp. 84–5 and Westermarck, 1929a, pp. 27–8)

Westermarck used Freud's works *Totem und Tabu* (the first German edition, 1913) and *Drei Abhandlungen zur Sexualtheorie* (the English translation of 1918) as sources in the revised edition of *HHM*. However, Westermarck had probably already heard about Freud's theories in the 1910s. In his memoirs, Westermarck refers to his seminar group that gathered at Box Hill to introduce and discuss Freud's 'new book' (perhaps *Totem und Tabu*) (MML, p. 257). Westermarck does not give exact dates for the seminar, but he talks about it in a chapter in which he discusses the years 1910–1914. He returned to his critique of psychoanalysis in 1934, in the wake of the Oedipus complex debate, when discussion on psychoanalysis had already markedly waned among anthropologists.

Freud's influence in the British intellectual community surfaced in the beginning of the 1920s when psychoanalysis became, for a short time, a fashionable topic among the British intelligentsia. Freud was not well known in England until the publication of his *Collected Papers* started in 1924–1925. Malinowski became a Freudian revisionist in England in the 1920s. He tried to apply psychoanalytic concepts and theory to anthropology, and he reformulated psychoanalytic paradigms on the basis of anthropological knowledge (see Malinowski, 1923, 1924).[6]

The revision of psychoanalysis

In his revision of psychoanalysis, Malinowski concentrated on a more systematic comparison of biological and sociological effects. In his *Sex and Repression in Savage Society*, Malinowski (1927b) questioned the universality of the Oedipus complex and identified culturally specific familial complexes. For example in a matrilineal society, the essentially patriarchal Oedipus complex is modified, the relative advantages of 'mother-right' being 'perhaps on the whole greater than those of father-right' (Malinowski, 1927b, p. 273). Malinowski (1924) had transferred the abstract concept, 'complex', from the framework of psychoanalysis to an anthropological framework and a social psychological one (as developed by Shand and McDougall).[7]

An orthodox Freudian, Ernest Jones commented on Malinowski's revisionist attempts. Jones respected Malinowski as a skilful field researcher, but considered his genetic interpretation false. Jones's school of psychoanalysis believed that the Oedipus complex was a universal foundation of the human culture; it was not a familial complex that varied according to family structure and cultural form. Jones (1925) identified the matriarchal social system and the ignorance of physical fatherhood among 'savages' as defences against original oedipal tendencies.

The dispute between Malinowski and Jones was the climax among anthropologists of the dispute over the Oedipus complex. It has usually been forgotten that Jones, Freud's loyal follower, had begun to criticise anthropological theories of the family as early as 1922, when he evaluated the revised edition of Westermarck's *HHM* as psychoanalytically untenable and absurd:

> Over and over again on reading the book one comes across a series of sterile speculations about some obscure problem the answer to which would be given at once by some knowledge of the unconscious, i.e. primitive, mind such as is afforded by psychoanalytic research. [. . .]
>
> It will not be long before the results of these researches, which he [Westermarck] now despises, will belong to the commonplaces of general scientific information, when many of the profitless speculations to which the author devotes so much attention will have been superseded. In spite of these strictures, however, we must repeat our opinion that his work remains a massive

> store-house of valuable information, one quite indispensable for all students of the subject.
>
> (Jones, 1922, pp. 250, 252)

In *HHM*, Westermarck had referred to Freud only in passing, but this was enough to convince Jones that the book was outdated.[8] Westermarck was bothered by this opposition from a spokesman for the orthodox psychoanalysts. In *A Short History of Marriage* (1926b), Westermarck commented on Jones's arguments and explained his main points of disagreement with the psychoanalysts:

> According to Dr. Jones, psycho-analysis points out that the strong and universal inclination towards incest really exists, 'only that for the most part it is repressed in the unconscious.' But *why has it been repressed?* I have found no satisfactory answer to this all-important question in the writings of psychoanalysts; and it is certainly a most unjustified claim to say that psycho-analysis has thrown 'a flood of light' on 'the almost universal horror of incest, and the extraordinarily complicated and fierce laws that have been devised in the most varied parts of the world with the object of preventing it.' For what reason have those complicated and fierce laws been devised; and how have laws in this particular case, but not in others, been able almost completely to repress an innate sexual desire, especially if it was once universally felt? These are the questions which should be answered before the existence of a conscious incestuous desire in primitive men is postulated in consequence of conclusions drawn from psycho-analytical investigations relating to a few thousand civilised individuals, mostly neurotic, and from myths and cosmogonies which allow of various interpretations.
>
> (SHM, p. 85)

Westermarck studied psychoanalytic literature more extensively in the beginning of the 1930s, when he was preparing to write an essay on the Oedipus complex, which was his major attack on psychoanalysis.[9] Before he managed to finish and publish it, however, Malinowski had done his work on revising psychoanalysis. Malinowski had adopted his view of the family as the core of social organization from Westermarck. Unlike Jones, Malinowski regarded Westermarck's book on marriage as a milestone of the sociological method and thinking. He respected the fact that Westermarck had not set up a hypothetical construction to serve as the narrow windows through which readers would have to stare at reality, seeing only as much as they allow (Malinowski, 1922b, p. 129).

Neither Westermarck nor Malinowski tried to construct a broad and systematic theory dealing with the relationship between the Oedipus complex and human culture. For psychoanalysts, the Oedipus complex was a (psychic) 'reality' (a cross between fantasy and reality) but not a (concrete, sociological, anthropological) 'fact'. For Westermarck and Malinowski, the Oedipus complex was a cultural and a pragmatic fact (familial complex) which could be anthropologically tested. Whereas Malinowski revised the concept of the Oedipus complex on the basis of anthropological and socioeconomic observations, Westermarck completely abandoned this psychoanalytic concept on the basis of his social anthropological data.

By 1927 Malinowski had reached a turning point in his revision of psychoanalysis. In the preface of *Sex and Repression in Savage Society*, he did not count himself among the supporters of psychoanalytic theory. Nevertheless, he did admit that psychoanalysis had offered him ideas and valuable guidance in reference to the 'unofficial' sides of human psychology, especially ambivalent emotional relationships, expressions of sexuality and the animal aspects of human nature:

> The open treatment of sex and of various shameful meannesses and vanities in man – the very thing for which psycho-analysis is most hated and reviled [. . .] should endear psycho-analysis above all to the student of man. [. . .] Man is an animal, and, as such, at times unclean, and the honest anthropologist has to face this fact.
>
> (Malinowski, 1927b, pp. vii–viii)

Malinowski did not consider himself to be in total opposition to psychoanalysis. Nevertheless, neither he nor Westermarck accepted the connection between the Oedipus complex and infantile sexuality and pregenital fantasies. Both scholars examined infantile sexuality in light of adult genital sexuality. Furthermore, for them both, the predominance of genital sexuality and heterosexual coupling, as well as monogamy and jealousy, was a natural guarantee of the continuity of culture and social order (Westermarck also included homosexuality among the culturally acceptable variations of sexuality). Even though they considered the psychoanalytic theory of instinct and the psychoanalytic interpretation of symbol formation to be overly metaphysical, both Westermarck and Malinowski emphasised, in accordance with Freud, the cultivating and refining influence that cultural experience had on instincts and drives (or 'needs'). According to Westermarck and Malinowski, the conclusions Freud drew about social institutions being based on instincts were unanthropological and removed from the 'realism' of fieldwork, that is, of 'concrete' observations on culture. Nevertheless, they both believed that research on sexuality and social research could support each other. The interface between anthropology and sexuality research could be remoulded.

Especially Malinowski, so to speak, needed the needs, that is, his functionalism used a psychological explanatory framework that held customs as equivalent to unconscious emotional forces (not as motives), and physiological needs as driving forces behind instincts, sentiments and emotions. For Malinowski, the 'nuclear family complex' was 'a functional formation dependent upon the structure and upon the culture of a society' (Malinowski, 1922b, p. 143) and 'a configuration of sentiments typical in a patriarchal or matriarchal society' (Malinowski, 1922b, p. 178; cf. also Fortes, [1957] 1968, pp. 168–72). Such a complex is also available for ethnographic field observation that is derived from customs and other behaviours giving empirically observable grounds rather than solely indirect (symbolic) derivatives from presumed unconscious and intrapsychic oedipal conflicts.

The Malinowskian nuclear family complex with family organizations and sentiments was tied to the primacy of culture (law, restrictions, regulations, power relations) over instinct (passions, desires) and to the ways of handling 'two main perils of humanity: the tendency to incest and the revolt against authority'. Freud's writings inspired Malinowski both to revise psychoanalysis and to clarify his own views on incest vis-à-vis Westermarck's ideas:

> We [i.e., Malinowski] differ also from Freud in that we cannot accept incest as due to the innate behaviour of the infant. From Westermarck we differ in so far as the aversion to incest does not appear to us as the natural impulse, a simple tendency not to cohabit with persons living in the same house from infancy, but rather as a complex scheme of cultural reactions.
>
> (Malinowski, 1927b, p. 252)[10]

This definition has behaviourist overtones, which later came to dominate Malinowski's work at the end of his career, when he became acquainted with the behaviourists at Yale University. In the mid-1930s Malinowski (1935, p. 236) even remarked that 'in one way the whole substance of my theory of culture [. . .] consists in reducing Durkheimian theory to terms of Behaviouristic psychology'. Psychoanalysis went out of fashion among British social anthropologists by the end of the 1930s. Malinowski proposed cooperation between anthropology and psychoanalysis in the last chapter of *Sex and Repression in Savage Society*,[11] but this suggestion did not meet with great enthusiasm. This book by Malinowski, as well as his *The Sexual Life of Savages in North-Western Melanesia* (1929), convinced Westermarck that the psychoanalytic method was unsound. He believed that it was necessary to attack that cornerstone of psychoanalysis, the Oedipus complex, with even more determination than Malinowski had done.[12]

Although Westermarck completely approved of the kind of sexuality research (based on Havelock Ellis's ideas) and psychology of sentiments and complexes (following Shand) that Malinowski practised, he did not follow Malinowski in his 'flirtations' with psychoanalysis. Westermarck confessed in a letter to Lagerborg that he regarded *Sex and Repression in Savage Society* as suspect in its shifting attitude toward psychoanalysis and the Oedipus complex:

> I resent such carelessness in the composition of a book [. . .]. His statement that there can be no Oedipus complex in a matriarchal society is absurd, if it implies that it can exist in a patriarchal society. In any case, it is the boy's father who has intercourse with his mother, and it is of the father the boy is jealous.
>
> (Westermarck to Lagerborg, 21 March 1934,
> ÅAB: EW, Box 18, number 2953)

Later Westermarck added with relief, 'he [Malinowski] no longer believes in the infantile desire for incest; his Freudianism was a short-term flirtation' (Westermarck to Lagerborg, 10 May 1934; cited in Lagerborg, 1951, p. 241). The 'flirtation', however, was not as short-term as Westermarck would have wished. Malinowski continued to follow with interest the development of psychoanalysis; he personally knew Princess Marie Bonaparte, J.C. Flügel, Otto Rank and Wilhelm Reich.[13] As late as 1938, Freud himself sent a friendly message to Malinowski:

> I came to know so much about your person and your work that I will be very glad to meet you. I am pleasantly surprised that you should call yourself an adherer of Psychoanalysis, as hitherto I have been more aware of your opposition and contradictions to our views.[14]

Westermarck and Freud

Unlike Malinowski, Westermarck always considered psychoanalysis loathsome and totally unsuited for use in his own field of research. Nevertheless, he did not attack Freud's therapeutic practices, which he also points out to Lagerborg in a letter:

> As you have noticed, I do not at all intrude on his (Freud's) psychoanalysis of his patients, but criticize the anthropological support he thinks he can give to the universality of the complex. Here I move within my own field, of which Freud is rather ignorant.
>
> (Westermarck to Lagerborg, 31 July 1934, ÅAB: RL, Box 88, number 12142)[15]

Even though Freud clearly considered Westermarck's theory to constitute the antithesis of his own, and even though Westermarck later opposed psychoanalysts' theories, Freud and Westermarck shared a common foundation: the heritage of Enlightenment philosophy, the theory of evolution and atheism.

Freud had read three of Westermarck's books in German translation. In his own private library, he had *Geschichte der menschlichen Ehe* (second edition, 1902), *Sexualfragen* (1909; a collection of passages from *ODMI* printed as a special edition) and *Ursprung und Entwicklung der Moralbegriffe* (1907–1909).[16] In *Totem und Tabu*, Freud only briefly referred to Westermarck. He sided with Frazer against Westermarck in the debate on the incest taboo (see Freud, [1912–1913] 1978, pp. 148–9). Frazer (1910, pp. 71–169) had argued that there would be no need for a universal, stern prohibition of incest if there were a natural, instinctive aversion to commit incestuous acts. For Frazer, at the core of the incest prohibition was the institution and the widening practice of exogamy for the benefit of the community. Freud did not believe that Westermarck's explanation for the primitive and instinctive fear of the dead was sufficient; instead, he brought in as the primary explanation ideas from the psychoanalytic theory of neuroses: obsessive-compulsive symptoms, projections and emotional ambivalence (Frazer, 1910, pp. 74–8). Freud also disagreed with Westermarck's criticism of the research on group marriage (Frazer, 1910, p. 12). On the whole, however, Freud never engaged in a detailed criticism of Westermarck's work.

In *Totem und Tabu*, Freud had categorically abandoned Westermarck's idea of an 'inborn aversion' on the basis of Frazer's arguments and his own psychoanalytic experience (Frazer, 1910, p. 151).[17] Freud stated that he wrote this anthropological work as a response to Wundt's *Völkerpsychologie* and to Jung's attempts to solve problems of individual psychology with the help of 'folk psychological' material. Freud believed that psychoanalysis thoroughly served the purpose of examining taboos. In the examination of totemism, however, he acknowledged the difficulties of reconstruction, although he believed that his theory came close to the truth (Frazer, 1910, pp. 3–4).

Freud was influenced by evolutionist anthropological ideas through his reading of anthropological literature. In *Totem und Tabu*, he presented a speculative developmental arc from nature to culture: from the patricide that took place in the primal horde ('Im Anfang war die Tat'; Frazer, 1910, p. 194) via feelings of guilt to totemic practices, and finally on to patriarchal laws and exogamy. Freud was aware of methodological problems of anthropological research. According to Freud, the verification of an 'original state of affairs' was always a 'construction'. The researcher interprets deeds, thoughts and feelings according to his or her own 'psychic constellation'; the researcher forms his or her own category of 'the primitive' (Frazer, 1910, p. 125, note 1).

The attempts by Freud and other psychoanalysts to develop psychoanalytic anthropology were met with strong resistance. Criticism by Malinowski and Westermarck in the Oedipus complex debate was, especially among American anthropologists, complemented by writings criticising Freud for relying on outdated evolutionist ideas, speculative historical reconstruction (the tale of the 'primal horde'), Lamarckism and the recapitulation thesis (the identification of the mental life of children, neurotics and 'savages': the identification of the 'present' and the 'ancient').[18] The 'universalist' psychoanalysis and the 'relativist' anthropology were increasingly interpreted as incompatible opposites.

Westermarck's contribution to the Oedipus complex debate

Towards the end of the 1920s, it became important for Westermarck to write his own estimate of the position of psychoanalysis, and of the Oedipus complex in particular. In his *Short History of Marriage* (SHM, 1926b), he had already given a reply to psychoanalytic criticism of his theory of exogamy. Finally, *Three Essays on Sex and Marriage*, published in 1934, included Westermarck's defence of his academic integrity against attacks by Robert Briffault, but also a relatively late polemic against Freud. At that point, the debate on the Oedipus complex was already settled in the Anglophone world.[19] Westermarck (1934a) also published his essay on the Oedipus complex in Swedish.[20] In the 1930s, Westermarck was one of the few anthropologists who had studied Freud's publications in German and who had closely examined some of the core hypotheses of psychoanalysis.

In the essay on the Oedipus complex, Westermarck did not so much touch upon Freud's connections with the theory of evolution, which reflects the fact that both Freud and Westermarck approached the historical roots of the aversion against incest from a bio-psycho-evolutionary angle. They both also discussed the effects of childhood experiences and familial socialization on the formation of an incest barrier, an aversion against incest. However, Freud emphasised the intrapsychic aspects of the aversion against incest: dynamism, the unconscious, repression and ambivalence. In the psychoanalytic interpretation, the intrapsychic 'aversion' is twofold: incest appears as both repulsive and intriguing. Freud also referred to the central role of psychosexual development (especially between the ages of two and six) in the psychic construction of the 'aversion' and the dynamics of incest. The intrapsychic 'aversion' foregrounded by the psychoanalytic theory can become projected and externalised onto the social arena, and symbolically represented by variable, culturally coded taboos and regulations that anthropologists and other researchers study. Psychoanalytic research on cultural products requires the existence of fundamental, primitive and 'irrational' fantasies, as well as the existence of a universal human psychic 'apparatus'.

In his essay on the Oedipus complex, Westermarck opposed the psychoanalytic assumptions regarding the unconscious, repression, infantile sexuality, overdetermination, ambivalence and the interpretation of dreams. In the essay, Westermarck only defended the psychology of consciousness and accepted only the 'physiological' unconscious, not the Freudian one:

> As for me, I cannot find any other clear definition of psychic processes and psychic dispositions than one that requires consciousness as the central characteristic. There are naturally unconscious causes for them: they have physiological correlates which can be considered as causes. However, the Freudian unconscious is, as we have seen, something completely other than the mere absence of consciousness.
>
> (Westermarck, 1934a, p. 10)

Even though Westermarck, in his essay, mainly criticised the attempts to apply psychoanalysis to anthropology, he also expressed his doubts about the validity of psychoanalytic clinical observations. He based his doubts on the unreliability of memories, the intervening role of the analyst, the arbitrary and sexually tinged interpretation of symbols, the suggestibility of children and excessive generalizations from the mental life of neurotics (Westermarck, 1934a, pp. 12–14, 24–5, 28–31; see also Westermarck's [1928] observations on Moroccan beliefs related to sexual matters). Westermarck's criticism of the psychoanalytic method was largely founded on William Stern's and William McDougall's comments and included few original arguments. Based on his research in Morocco, Westermarck was convinced that the meanings given to dreams were usually based on similarity or contrast of associations, not on the dynamics of the disguised fulfilment of repressed wishes (see Westermarck, 1934a, p. 30; see also RBM II, p. 49).

Westermarck did not think that psychoanalytic theory had reformed or expanded the concept of 'sexuality', compared to Plato's use of the concept of 'eros'. Westermarck was absolutely opposed to Freud's argument of the universality of infantile sexuality and sided with Albert Moll (e.g., 1908, p. 13) in classifying infantile sexuality, described by Freud, as an abnormal phenomenon (Westermarck, 1934a, pp. 19–20, 36, 40–1).[21] Nevertheless, Westermarck did not deny the existence of incestuous relationships; he himself had collected many descriptions of such relationships (Westermarck, 1934a, p. 47). Westermarck considered it evident that parents' attitudes toward their children were based on their own sexual tendencies, but he did not speculate on the early developmental foundation of these tendencies. He also regarded as unfounded psychoanalysts' claims on the remnants of primitive incest (for example the marriage rules of regal families, levirate, sororate, group marriage and *jus primae noctis*) (Westermarck, 1934a, pp. 57–60).

The 'elasticity' of the psychoanalytic theory against counterarguments made Westermarck remark: 'If there exists evidence of primitive incest both in those cases where incest is accepted and in those cases where it is forbidden, it is fairly certain that it has once been a common practice all over' (Westermarck, 1934a, p. 67).[22] Because Westermarck doubted the scientific importance of the psychoanalytic concept of ambivalence, he accused psychoanalysts of reaching fantastic and far-fetched interpretations when they looked for traces of psychodynamically ambivalent attitudes in customs, beliefs and myths.

Westermarck disparagingly cited the pioneer of psychoanalytic anthropology, Róheim, and his observations on Central Australian children (Westermarck, 1934a, pp. 25–8). Róheim claimed to have encountered Freudian infantile sexuality and the Oedipus complex during his fieldwork, but he had not found the latency period (for Róheim's fieldwork report, see Róheim, 1932). Róheim formulated an ontogenetic theory, claiming that cultures were based on specific childhood traumas. His idea was that cultural variations could be traced to the solutions of libidinal conflicts: the basis of culture and the universal Oedipus complex was the prolonged human childhood and dependency and the early development of sexuality.[23] According to Westermarck, Róheim had inflated expectations as to the worth of his psychoanalytic 'findings' and of 'new anthropology' (Westermarck, 1934a, p. 28). In emphasising the determinacy of childhood experiences and the infantile and irrational bases of adulthood and institutions, Róheim also denied the existence of rational means of adaptation, formed in the process of growing up. On the other hand, many anthropologists, including Westermarck, had hardly any direct connections with research on childhood development or child-rearing practices. Therefore, they unduly emphasised the autonomous rationality of adulthood and the adult maturity of human institutions.

Westermarck (Westermarck, 1934a, p. 80; see also Róheim, 1932, pp. 35, 91, 98) believed that the fact that a lack of sexual interest during the latency period had not been observed among primitive peoples told against Freud's theory on sexuality. However, Westermarck failed to mention that, according to Freud, latency was not a psychosexual stage of development proper, but, rather, a consequence of cultural inhibition of sexuality. Westermarck (1934a, pp. 80–2) also claimed that primitive people rarely threatened their children with corporeal punishment and that the symbolic correlation between initiation rites (for example, involving circumcision or the removal of teeth) and castration remained obscure.[24] He thought that the threat of castration was connected with 'realistic' and conscious events. He was not convinced by the psychoanalyst's explanations of a web of fantasies that gets intertwined with the threat of castration.

Westermarck did not believe that the psychoanalytic interpretation of the motives and symbols in myths and legends revealed hidden wishes and fantasies; rather, it constituted a psychoanalytic mythology. Myths and legends (for example, *Oedipus Rex*) had undoubtedly contributed a great deal of material to psychoanalytic research. They had also shaped psychoanalysis in its early stages. Nevertheless, Westermarck disagreed with the Freudian interpretation of the Oedipus myth:

> It has been claimed that the tendencies that form the so-called Oedipus complex are most directly and most exhaustingly revealed in this legend. I cannot see how this statement can be defended. Not only did Oedipus unknowingly become his father's murderer and his mother's husband; not even in his unconscious could he have loved his mother or hated his father, because the former emotion which caused the latter emotion, presumably does not exist among children under the age of two, and Oedipus was taken away from his parents right after – according to Sophocles only three days after – his birth.
> (Westermarck, 1934a, p. 75)

If Oedipus had followed Freud's theory, his love and hatred, according to Westermarck's (Westermarck, 1934a, pp. 76, 110) interpretation, would have been directed toward his foster parents, not Jocasta and Laius. In Westermarck's view, wishes and fantasies must submit to the limits set by reality. He felt the attempts, arising from the oedipal situation, to deny the incest taboo and simultaneously to idealise and devalue the parents, to be a kind of fantasising that he found alien. These attempts were connected by psychoanalysts with the 'family romance phantasy'.[25] Psychoanalysis did not fulfil its task of resolving myths; instead, it became tangled in its own myths.

Freud ([1912–1913] 1978, p. 149) found peculiar the idea that inborn aversion against sexual relations (proposed by Westermarck) should prevail primarily among people who have grown up together, and only secondarily (through the law of association) among blood relatives – even though inborn aversion is supposed to be the psychic counterpart of the biological fact that incest is harmful for the species. Westermarck (1934a, pp. 110–11) replied to Freud's argument by remarking that, analogously, psychoanalysts were hard pressed to prove why and how the love a child feels toward those who rear it becomes an incestuous drive toward the parents (blood relatives) and what generates the repression of this incestuous drive. Furthermore, the psychoanalytic theory could not deduce from the Oedipus complex the aversion that the parents feel against sexual relations with their children.

According to the psychoanalytic theory, parents are the primary objects for the child's love. Childhood relationships between siblings usually do not turn erotic to the same degree; instead, they are marked by jealousy, hostility and competition, until they (usually) become balanced and socially reciprocal. On the other hand, the incest prohibition is not as strong between siblings as it is between parents and their children. In Freud's analysis, the focus was on the mother–son relationship at the expense of father–daughter and the brother–sister relationships. In addition, the nuclear family remained in the margins compared with the clan-totemic exogamous unit. Westermarck did not examine the structure and functions of different dyads, either; nor did he compare dyads.

One contribution of Westermarck's article on the Oedipus complex was that it summarised the discussion on the topic. Westermarck clarified the deficiencies of psychoanalytic anthropology, of which most psychoanalysts were aware. He paid particular attention to Freud's problems in forming a clear opinion on the role that heritability and education played in the creation and maintenance of the incest barrier. Freud was not able to give a satisfying explanation as to how incest taboos had been transferred from the past to the present. The problematic point in Freud's 'tale' was how the guilt of patricide could have been preserved in an individual without knowledge of the original deed. Freud's ([1912–1913] 1978, p. 190) partial answer was that 'psychic dispositions' were inherited but they needed a trigger in an individual's life to actualise themselves.[26]

Freud's and Westermarck's statements on this point were not mutually exclusive; Westermarck's 'inborn aversion' was also a kind of inherited 'disposition'. Psychoanalytic interpretations of the connections between natural selection and incest were, on the whole, inconsistent. Westermarck regarded it as illogical that psychoanalytic literature proposed a two-fold role for natural selection; it both advanced incest and led to its prohibition. Westermarck's (1934a, pp. 115, 125–6; see also Flügel, [1921] 1929, p. 207) own view was that the only way for natural selection to work towards the survival of the species was by eliminating harmful and destructive tendencies of sexuality and preserving its useful variations.[27]

Freud's tale of the primal horde was, for Westermarck, an additional psychoanalytic myth. He did not find support in ethnological literature for the real occurrence of the crime that Freud presented as the origin of social order, moral rules and religion. Furthermore, ethnological sources did not offer proof for the close connection that Freud saw between totemism and exogamy. There existed no evidence of the universality of totemism. Westermarck sceptically attempted to shake the foundation of Freud's tale: 'For my part, I cannot see that psychoanalysis has shed even one ray of light on the highly important question of *why* incest has been prohibited' (Westermarck, 1934a, p. 132).

In her response to Westermarck's essay on the Oedipus complex, Brenda Z. Seligman (1935, pp. 75–93) posited that researchers may never be able to explain the original reasons for the emergence and internalization of a cultural regulation like the incest taboo and proposed that researchers content themselves with correlating it with other cultural phenomena. Even if the Oedipus complex were universal, it would not have to be culturally uniform: the structure, strength and solutions of the Oedipus complex varied between different communities (Spiro, 1982, p. 163).[28] Anne Parsons has remarked that the 'original' debate on the Oedipus complex has lost its significance:

> For the original question of whether the Oedipus complex is universal or not, we would sum up by saying that it is no longer very meaningful in that particular form. The more important contemporary question would rather be: what is

> the possible range within which culture can utilize and elaborate the instinctually given human potentialities, and what are the psychologically given limits of this range.
>
> (Parsons, 1964, p. 328)

Westermarck (1934a, p. 161) left it to psychoanalysts to study the influence of the Oedipus complex on the life of the (neurotic) individual, and to sociologists to study the assumption of the universality of the Oedipus complex, as well as the effect the complex had on the development of civilization. In retrospect, Westermarck's own theories were not faultless. He did not clearly identify the instinctive basis of the inborn aversion against incest, nor did he clarify the connection between the incest taboo and marriage regulations (see Antfolk and Wolf, this volume). In addition, his model presented no explanation for the existence of actual incestuous relationships.

Westermarck had apparently expected psychoanalysts to criticise his essay severely. In the spring of 1935, he wrote a letter to Lagerborg (7 April 1935, ÅAB: EW, Box 18, number 29561) about the lack of response: 'Freudians keep silent, often the safest method for the attacked party'. The silence, however, was not complete. One of the leading British Freudians, J. C. Flügel, whom Westermarck (1934a, p. 125) had described as a 'moderate Freudian', published a review of Westermarck's criticism of the Oedipus complex in 1935.[29] Flügel's review followed the tenets of psychoanalysis, but he did not deny the worth of Westermarck's argument; Flügel recognised Westermarck's detailed attention to the gaps in psychoanalytic anthropology. Flügel also valued the fact that Westermarck had pointed the way to more specific research topics. According to Flügel (see 1935, p. 383), these topics should include different forms of totemism, the shifting relations between totemism and exogamy and the conditions for the appearance of repressions and aversions in legal prohibitions, social conventions and sanctions based on supernatural beliefs.

Westermarck's attack against psychoanalysis had been favourably received among anthropologists,[30] and ecclesiastical circles welcomed his polemic against Freud (see Lagerborg, 1951, pp. 244–5). At the end of their careers both Freud and Westermarck returned to the analysis of religion. Even late in the 1930s, Freud held on to the highly speculative ideas of Robertson Smith, according to which totemism was the earliest form of religion and the totemic feast was a model for religious altar offerings, including the Christian communion. In his last essay on Moses, Freud ([1939] 1978, p. 390) defended himself against his anthropological critics saying that he had the right to use ethnological literature for psychoanalytic purposes.[31] He reserved the right to study even the psychology of religious phenomena.

Even though Westermarck also studied religion and published his conclusions in his last book, *Christianity and Morals* (1939), he did not apply an 'origin and development' approach or a psychological method to the study of religion. Many researchers had already written on the origins and development of religion. Thus, Westermarck's task became to trace the origins and development of marriage and morality. He did not believe that reflections on ancient religion supported the

Oedipus complex hypothesis. The totemic feast was, according to him, neither a reproduction nor a reminder of patricide and cannibalism because social institutions and morals had not been based on such a crime (Westermarck, 1934a, p. 145). In psychoanalytic theory, the attempts of the guilt-ridden sons to propitiate the father-surrogates had been given the tone of a mystical revelation. Freud had trespassed on an area that Westermarck considered unsuitable for psychoanalysis.

Notes

1 See Westermarck (1925, pp. 27–8): 'Another question had actually occupied my thoughts: how shall we explain the veil of discretion and modesty surrounding sexual life, in contrast to other manifestations of human nature? I have later come to discover that Church Father Augustine had been pondering over the same problem. To find an answer to the question I started to study such customs and manners of various people that are in some way connected to the subject and was, in this way, without any special inclination, driven into the haven of married bliss.' See also Westermarck (1921d) and MML, p. 69.

2 This chapter is based on revised extracts (mainly subchapter 6.3) from my work *Multiple Origins: Edward Westermarck in Search of Mankind* (Ihanus, 1999). The translations from the Swedish sources are mine.

3 Freud's work was originally published as four essays in *Imago* in 1912–1913. Freud actively supported Friedrich Salomo Krauss in the publication of the yearbook *Anthropophyteia* (1904–1913) which presented folklore and ethnological materials on sexual life (compare Reichmayr, 1994, pp. 29–51; Reichmayr, 1995, p. 29; also Wallace, 1983, p. 47).

4 See, e.g., Róheim (1917, 1918, 1919); See also Ihanus (1998).

5 See, e.g., Róheim (1932). The fieldwork was financed by Princess Marie Bonaparte and prepared with the help of Freud, Ferenczi and Kovács.

6 Havelock Ellis excitedly greeted Malinowski's revision of psychoanalysis in a letter to Malinowski (8 November 1923; BMC):

> Your letter in *Nature* on psycho-analysis was of much interest to me & I sympathize with Your sympathetic & yet judicial & discriminate attitude. I have been in touch with Freud & his work ever since he published his first work in 1896 [*sic*], and – though on different grounds – my attitude very much compounds to Yours.

In the same period, a discussion about the relationship between psychoanalysis, psychology, and anthropology was also going on at the Royal Anthropological Institute (see Jones, 1924; Seligman, 1924; see also Pear, 1960).

7 Malinowski (1927b, p. 75) confessed that his interpretation of 'complex' differed from psychoanalysts' use of the term. Malinowski talked about 'complexes' and 'sentiments' in the same manner as Shand who, according to Malinowski (Malinowski, 1927b, p. 241) plotted a concrete psychology of sentiments:

> Had the founders of psycho-analysis known Shand's contribution, they might have avoided a number of metaphysical pitfalls, realized that instinct is a part of human sentiments and not a metaphysical entity, and given us a much less mystical and more concrete psychology of the unconscious.

According to Malinowski (Malinowski, 1927b, pp. 241–2), Freud's contribution to the 'theory of sentiments' was to emphasise the role of family and repression in the formation of 'sentiments.'

Shand (1922, p. 126) himself preferred the term 'system' to the term 'complex' when dealing with 'the normal mind.' Shand (Shand, 1922, p. 128) made a distinction between 'sentiment' and 'emotion.' The former is 'a mental system which is capable of adapting itself emotionally to the changing situations of its object, and of persisting indefinitely in correspondence with the duration of its object without becoming morbid'; the latter is 'a mental system which is capable of adapting itself to only one kind of situation, and when it persists beyond this system tends to become morbid.' – Shand's *The Foundations of Character* (1914) was highly esteemed by both Westermarck and Malinowski.

8 It irritated Jones (1922, p. 252) that Westermarck (HHM II, p. 204) had called Jung the most important 'disciple of Freud's': 'We would also inform him [Westermarck] that Dr. Jung, so far from being a "disciple of Freud's", is one of Freud's chief opponents, so that he might just as well quote his own opinion as Dr. Jung's.'

9 Westermarck mentions his intentions to write a separate essay on the Oedipus complex in a letter to Lagerborg (16 December 1928, ÅAB: RL, Box 88, number 12112).

10 Because incest was incompatible with the family organization and its function as the medium of transmitting culture, Malinowski (1927b, p. 252) reasoned that if incest were permitted, 'the fundamental pattern of all social bonds, the normal relation of the child to the mother and father would be destroyed.' See also Fortes (1957/1968), p. 172.

11 Malinowski (1927b, p. 279):

> The anthropologist has received some help from the psychoanalytic school and it would be a great pity if the exponents of this latter refused to collaborate, to accept what is offered in good faith from a field where, after all, they cannot be at home.

Later Malinowski (see [1944] 1969, pp. 22–3) remarked that Freud had been able to shake the Western taboo of 'indecency,' but he added: 'The approval of psychoanalysis does not in any way detract from the great importance which behaviorism promises to acquire as the basic psychology for the study of social and cultural processes.'

12 In the summer term of 1929, the themes of Westermarck's seminar at the London School of Economics (LSE) touched family, marriage, the Oedipus complex and psychoanalysis. In a letter to Lagerborg (17 June 1929, ÅAB: EW, Box 18, number 2935), Westermarck described his seminar and praised Malinowski, who used to attend Westermarck's seminars:

> The first lecture was given by Malinowski, who stood by me all the time; he has recently published a great work on the sexual life of savages, with a preface by Havelock Ellis, which has attracted a lot of attention. [. . .] A semi-Freudian talked about the 'Oedipus complex,' and both the lecture and the discussion following it, in which a practicing psychoanalyst participated, made me even more convinced of the oddities of the theory.

13 Wilhelm Reich, for example, in his letter to Elsa Lindenberg (28 February 1935; cited in Reich, 1994, p. 33), talks about Malinowski's support: 'Today I received a very kind letter from Malinowski. He is very enthusiastic about the *Invasion [of compulsory sex-morality]*. He is singing its praises everywhere and is using my theories in his lectures' (emphasis in original).

14 Freud's letter to Malinowski on 22 June 1938 (BMC). Freud's note is from the period when Malinowski had sent Freud information about the meeting of the Sociological Society (17 June 1938; see Jones, 1957, p. 250). In his late years, Freud was prepared to include Malinowski among the supporters of psychoanalysis, and similar rumors were found in the press. In Italy, in 1938, Malinowski had read an article in *Corriere Della Sera* where he, Freud and Adler were described as 'Jewish sexologists.' Malinowski considered this a public circumcision '*in effigie*,' in which the writer of the article had made three mistakes in a single sentence. Malinowski wrote in a letter to Richard Thurnwald (10 September 1938, BMP): 'For Adler is not a sexologist, and, for myself, I hope, I am neither Jew nor sexologist.'

15 Even though Westermarck abandoned psychoanalysis, he did not personally dislike Freud. Westermarck remarks in a letter to Malinowski (17 May 1933, BMC): 'Freud is an honest man, whom you can respect, even when you thoroughly disagree with him.'

16 See *Concise Catalogue of Freud's Library* (s.d., Freud Museum, London). Westermarck's books were among those that were moved from Vienna to London.

17 In his own writings, Freud usually distinguished between the concepts 'Trieb' (drive) and 'Instinkt' (instinct). In the English translations of his texts, 'Trieb' often has mistakenly been translated as 'instinct.' Psychoanalytically, 'drive' ('Trieb') falls in the middle ground between the mental and the somatic. Impulses born within the organism get a psychic representation, which requires from the mind psychic work in relation to the body. Westermarck again used the word 'instinct' in English and the words 'drift' and 'instinkt' in Swedish, without making a distinction between them. From a psychoanalytic perspective, Westermarck mainly appears as a Darwinist researcher of sexuality who uses the theory of instincts.

18 For early American anthropological criticism of psychoanalysis, see Du Bois (1937); Kroeber (1920, 1939); Kroeber (1920, p. 55) left the door open for psychological anthropology:

> However much cultural anthropology may come to lean more on the historical instead of the psychological method, it can never ultimately free itself, nor should it wish to, from the psychology that underlies it. To this psychology the psychoanalytic movement initiated by Freud has made an indubitably significant contribution.

In the first British review of the English translation of *Totem und Tabu*, Marett (1920, pp. 205–6) called Freud's report a 'just-so story.' Marett admitted that Freud had offered anthropology a valuable method for analyzing ambivalent emotional states. However, according to Marett, the psychoanalytic interpretation did not help in better understanding the origins of exogamy and totemism, nor did it capture the characteristic traits of 'savage' sexuality.

19 Westermarck was aware that the discussion on the Oedipus complex had slowed down, as he remarks in a letter to his publisher, Macmillan (12 December 1933, MA, Add Ms 55156, folio 204): 'someone has told me that people are getting rather tired of the Oedipus complex.'

20 The citations in this chapter are from the Swedish version revised by Westermarck (1934a).

21 Toward the end of his life, Freud ([1938] 1978, pp. 74–5) believed that psychoanalysis had expanded the meaning of the concept of 'sexuality' with three central arguments: (i) Sexual life begins to appear soon after birth, not only in puberty; (ii) 'Sexual' and 'genital' should strictly be kept separate. 'Sexual' is more extensive than 'genital' and covers actions which are in no way linked to genitals; (iii) In sexual life, pleasure is found through the erogenous zones of the body. Later, the pleasure function serves the reproductive function. These two functions (pleasure and reproduction) are often not completely compatible.

Freud's theory of drives was removed from traditional perceptions according to which sexuality was a biological force looking for a natural object. For Freud, sexuality was not a single uniform sexual drive, but, instead, a complex dynamic entity, psychosexuality, which balanced, on the one hand, between the biological source and impulse of the drive and, on the other hand, between the goal of the drive and the psychic organization of the object. Freud's model of sexuality can be considered masculine and reproduction centered. However, within the model, it is possible to pay attention to other, marginally developing sexualities through the identification of a shared universal human developmental 'deviation,' an infantile 'perversity' and 'bisexuality.'

22 Westermarck repeated this statement in his last methodological comment (Westermarck, 1936, p. 235).

23 Róheim has also been criticised for failing to explain the potential for cultural change, although he talks about cultural evolution (see Bourguignon, 1979, p. 90). He does not specify the influence of 'external' factors, and he spins a theoretical vicious circle: personality generates from childhood trauma and the same trauma repeats itself in the following generation.

24 On the psychoanalytic interpretation of initiation rites, see Reik (1928/1931).

25 On the 'family romance phantasy,' see Freud ([1909] 1976). The standard English translation of the original *Familienroman* should actually be rendered more literally as 'family novel' (phantasy). According to psychoanalysts, a myth is not a fundamentally rational cultural product, even though it can be revised and rationalised, and even though it can, as a metaphysical system, offer explanations and interpretations. A myth, which is linked to the primary process, is 'timeless' and usually without 'negations.' According to psychoanalysts, the possible logical inconsistencies and paradoxes of a myth prove the close connection between myths and significant, hard-to-process psychic conflicts (cf. Hook, 1979, p. 286).

26 Freud (Freud, [1909] 1976, pp. 191–4) also mentioned that 'psychic' (not 'factual') realities lay behind neurotic feelings of guilt. The neurotic has been restrained from action; therefore, the thought is a substitute for the deed. The 'primitive' man's actions, on the other hand, are unrestrained and thus the thought is immediately replaced by a deed.

27 Badcock (1980, pp. 31–7) has defended the ambiguity of Freud's views on inheritance by remarking that Freud could not yet have accurate knowledge on natural selection and mechanisms of inheritance. See also Badcock (1994) on 'psychodarwinism.'

28 Cross-cultural psychology is rarely relativist. Psychologists have not formulated a complete relativist model of cognitive or emotional development. From the empirical point of view, adopting absolute relativism impedes psychological research and makes ethnography chaotic and complicated. In anthropology, relativism replaced Western arrogance and a hierarchical division into 'us' and 'them.' Relativism has, however, also produced 'cognitive apartheid': if 'we' cannot be better than 'they' within one world, let everyone live in their own world (see Sperber, 1985, p. 62; on relativism in anthropology and psychology in general, see Sperber (1985, pp. 35–49).

29 In 1935, Freud's close friend, M.D. Eder (1935, pp. 115–16), had also written a short presentation of Westermarck's earlier work (Westermarck, 1932b) and had classified Westermarck among the most skillful representatives of nineteenth-century rationalism who lacked the understanding of new psychology (psychoanalysis).

30 See, e.g., Rose (1934, p. 357): 'he [Westermarck] amuses himself and his readers at the expense of that bogey of the Freudians, the notorious "Oedipus-complex," which he lays, not only by the strong conjurations of his ethnological science, but with the scarcely less effective rites of a mordant humour. It has always been good magic to make fun, and never was it used to better purpose.'

31 Westermarck was not blind to the connection between religion and sexuality (for example, his examination of religious prostitution and celibacy in ODMI; see ODMI II, chapters XLI and XLII). For Robertson Smith's views on religion, see Smith ([1889] 1894).

4 The sensitiveness of the holy

Westermarck and Durkheim on society and religion

Olli Lagerspetz and Kirsti Suolinna

In connection with his research on popular religion in Morocco, Westermarck encountered ideas and practices that, according to him, would disprove Émile Durkheim's theory of holiness (*le sacré*). In particular, he was struck by what he called the sensitiveness of the holy: the fact that holiness (*baraka*) could be challenged and renegotiated. This fact, which Westermarck called 'the socially most important characteristic of the holy', attracted his interest still shortly before his death in 1939. In the present chapter, we explain why he saw it as a disconfirmation of Durkheim's theory.

Westermarck went to Morocco originally in 1898 in order to find concrete examples of how morality was influenced by religious and magical practices and ideas. He discovered a rich material of folklore and folk belief. However, the present chapter does not advance claims about reality in Morocco per se. We are interested in how Westermarck's meeting with Morocco caused him to adopt a more nuanced view on the social role of religion than his initial theoretical approach would suggest, although he still did believe that this newfound reality was compatible with his generally sentimentalist take on religion. In brief, Westermarck presents the holy as a more dynamic phenomenon than either Durkheim or Frazer would have it.

Durkheim (1858–1917) and Westermarck (1862–1939) were near contemporaries. Both saw their research as a contribution to the debate on the 'foundations' or 'origins' of morality and religion, at the time considered a philosophical *cause celèbre*. The debate mixed elements from moral or social philosophy and empirical anthropology. In the process, it became a breeding ground for the modern social sciences (see Lagerborg, 1953). One part of the battle was fought over the notion of 'the holy'.[1] Whereas Durkheim explained the concept of holiness in terms of a communally imposed system of rules, Westermarck thought its main source lay in the emotional experience of wonder, which was ultimately rooted in human and animal nature.

Durkheim dedicated his last great work to religion, connecting it to his general theory of social solidarity. *The Elementary Forms of the Religious Life* ([1912] 1968) had a lasting influence on the social sciences as a statement of his ideas on the social role of religion – indeed of his entire theory of society. It became a cornerstone of what came to be known as new anthropology, and generally of structuralist and functionalist approaches.

Westermarck put forward his views on holiness over a time span stretching from 1905 to the very last months of his life. He drew on his extensive fieldwork in Morocco. Early in the 1900s he arrived at theoretical formulations that were carried over almost unchanged to his later monographs on popular religion in Morocco. However, the monographs from 1914 onwards show a change of focus, in fact if not in theory. Westermarck's earlier theoretical formulations linger on, but his actual focus moves on from religious feeling to the social fabric that sustains holiness. His data are consequently of interest for the study of the political and social dimensions of religion. As Kenneth Brown (1982, p. 227) observes, Westermarck's field studies are valuable because their wealth of detail allows for reinterpretation also by those who do not share his theoretical views. We argue that a fruitful reanalysis of Westermarck's *Ritual and Belief in Morocco* is possible (also see Shankland, 2014b, p. 25).

Durkheim, on the other hand, insisted *in theory* on the need to investigate the social functions of religion, but due to his heavy reliance on material from supposedly homogeneous Australian tribal societies, he missed out on important features of such functions. In particular, we suggest he downplayed the fact that all societies are, to some extent, pluralistic and that ideas of holiness will hence be contested according to the needs and perspectives of different social agents. In contrast (and, one might say, paradoxically – considering Westermarck's resistance to functionalist theory), Westermarck's field results from Morocco highlight such aspects of the *social functions* of holiness.

Westermarck's early work on the holy

Westermarck presented his early views on the holy and the profane in a number of brief essays in the first decade of the 1900s (Westermarck, 1905, 1906b, 1907). He argued that the idea of the holy is based on emotions of awe and respect, a sense of 'the mysterious' (Westermarck, 1905, p. 21, cf. RBM I, p. 21). The formulation included in *The Origin and Development of the Moral Ideas*, in 1908, can be singled out as his final definition of religion. It is reiterated in his subsequent works, including *Ritual and Belief in Morocco*:

> Religion is a belief in and a regardful attitude towards a supernatural being, on whom man feels himself dependent, and to whose will he makes an appeal in his worship.
>
> (ODMI II, p. 582; RBM I, p. 32. Also see Westermarck, 1906b, p. 144)

This definition of religion is couched in terms of individual psychology. The distinction between religion and magic is based merely on the practitioner's attitude ('regardful' or otherwise), not on institutional differences between them. Magic is a technique for wringing concessions from the gods instead of appealing to them (Westermarck, 1905, p. 25; RBM I, pp. 17–34). In this context, Westermarck refers approvingly to James Frazer's anthropological classic, *The Golden*

Bough (Frazer, 1911, p. 220).[2] However, Frazer believed that magic and religion belonged to different stages of cultural development, a view that Westermarck contested (Frazer, 1913, pp. 304–5; Westermarck, 1905, pp. 28–9).

In addition to feeling, religion includes a theoretical element – that of religious belief – and a practical one – that of cult (Westermarck, 1905, p. 24; 1907, p. 9).[3] Religious belief and cult are attempts to articulate or rationalise the underlying emotion. Westermarck cites his Moroccan experiences. Unusual features of the natural landscape, such as large lone trees, grottoes, natural springs and conspicuous rocks on the seashore, are typically regarded as holy in Morocco (Westermarck, 1905, p. 21). Originally, these sites doubtlessly owe their holiness to their importance in the landscape. However, they are now without exception associated with specific Muslim (or sometimes Jewish) saints who supposedly stayed there or are buried there. The Islamic belief framework is superimposed on natural sites whose sacred status predates the arrival of Islam.

In sum, Westermarck asks why such and such *particular* objects and people are singled out as holy, and he provides an answer in terms of their aptitude to evoke feelings. Although Westermarck does discuss the relation of magic to 'social relations', he mainly thinks of the relation as one sided: of magic influencing social relations, not the other way around (Westermarck, 1906b).

Durkheim on the holy

Durkheim's account of religious belief agrees with Westermarck's on a number of points, but his fundamental take on religion is different. To a great extent, the conflict between them concerns the question of *what one should look for* in a study of religion and society. Durkheim's question is: Why do humans make a distinction at all between the holy and the profane? Durkheim takes exception to the view that human reactions to nature could, on their own, give rise to feelings of holiness. On the contrary, primitive men were far too well acquainted with natural phenomena to marvel at them (Durkheim, [1912] 1968, p. 84). More to the point, there is nothing in natural disorders and accidents as such to mark them out as sacred as opposed to merely surprising and unusual (Durkheim, [1912] 1968, pp. 28–9, 84–5). The idea of the sacred is *sui generis* and not reducible to the unexplained. The distinction between the holy and the profane is, according to Durkheim, universal. There will be differences between societies on how to classify specific objects, but the two main classes remain the same. Nowhere else in the history of human mind have there been two categories so radically distinct. Prohibition and taboo mark the boundaries between the profane and the holy, thereby structuring life for members of the community (Durkheim, [1912] 1968, pp. 127–33).

The disagreement between Durkheim and Westermarck on the role of natural phenomena is simply an outcome of their central difference concerning the role of religion. Durkheim sees its main role as that of regulating social relations. Feelings of holiness owe their existence to relations between *human beings*, not to relations between man and the physical environment. Natural objects enter the picture only as arbitrary focal points for socially imposed rules of conduct. A religion must, according to Durkheim, involve two elements: a community or a 'Church' and the categorisation of reality into holy and profane spheres. In brief:

> *A religion is a unified system of beliefs and practices related to sacred things, that is to say, things set apart and forbidden – beliefs and practices which unite into a single moral community called a Church, all those who adhere to them.*
>
> (Durkheim, [1912] 1968, p. 47, italics in the original)

The essence of holiness lies in the existence of a socially sanctioned system of prohibitions. The difference between religion and magic lies simply in the fact that the former, unlike the latter, has the social backing of a church.[4] Durkheim was not only making a limited point about the definition of a theological or anthropological term. Religion captures his interest because, for him, it is the meeting place for everything crucial for the relation between individual and society: external, anonymous, collective pressure on individual thinking and, as the source of this pressure, the 'collective consciousness' of the community. As Bronisław Malinowski would point out in his review of *The Elementary Forms of the Religious Life*,

> To Prof. Durkheim the religious is the social *par excellence*. The distinctive characters of social and religious phenomena practically coincide. The social is defined, in *Règles de la méthode sociologique*, by its 'exteriority to individual minds,' by its 'coercive action' upon individual minds; the religious, which is also 'external' to individual minds, by its 'obligatoriness'.
>
> (Malinowski, 1913, p. 525, original emphasis)

The role of empirical data

The argumentative force of Durkheim's conception of religion is largely based on conceptual analysis and stipulative definition. He does not think of it merely as applicable to specific primitive societies, but to society in general. Any recognisably social life must involve collective pressures, and these must be mediated by a system of beliefs sanctioned by a community of believers. It is perhaps best to see Durkheim's work as an analytic attempt to determine the *status* of moral and religious statements – to identify their logical subject. To classify a statement as moral or religious in character (rather than, say, magical) *is*, according to Durkheim, to say that it involves the implicit claim of being backed by a community.

However, this invites the question of what role is to be assigned to the empirical material that Durkheim used for illustration. Westermarck was irritated by the fact that Durkheim's data came almost entirely from the single case of Australian totemism. In a letter to his friend Rolf Lagerborg, he was curious to know how Durkheim was entitled to claim that his generalisations 'out of the blue' apply to all religions. 'By intuition, I presume', he ironically concluded, 'for Australian

Totemism cannot justifiably support all these general claims' (Westermarck to Lagerborg, 8 March 1914. ÅAB: RL, quoted in Lagerborg, 1951, p. 428). Westermarck (1936, p. 240) persisted in his critique of Durkheim still in 1936 in his last major public lecture, *Huxley Memorial Lecture*, with the title *Methods in Social Anthropology*. He argued that a comparative approach was necessary.

Westermarck's point is well taken, but the question is how this reflects on Durkheim's concerns. One may ask what kind of a generalisation is involved in Durkheim's theory. Inductive generalisation is out of the question, as Durkheim clearly concedes ([1912] 1968, pp. 94–5). But an individual case may be useful for developing a *deductive* argument. Durkheim treats Australian tribal communities, as it were, as *prototypes* of more complex societies. The very fact that makes them poorly representative – their assumedly 'homogeneous' and 'primitive and simple' organisation – is also what makes them appealing as illustrations (Durkheim, [1912] 1968, pp. 95–6). On the other hand, this invites the question of what is *missed* when (allegedly) simple and homogeneous societies are privileged as models for social theory. The model naturally tempts us to think that all societies, despite appearances, somehow essentially or at bottom *are* homogeneous and, connecting to this, that religion only has one function – that of creating stability.

We believe that Westermarck's studies of the concept of holiness in Morocco contribute to criticism of this kind. Westermarck himself said that 'the socially most important characteristic of the holy' was 'its extraordinary sensitivity to external influences' (Westermarck to Lagerborg, 18 April 1939, ÅAB: RL); 'There is not that impassable gulf between the holy and the profane which has been postulated by Durkheim' (RBM I, pp. 146–7; cf. Westermarck, 1916, pp. 75–6, Westermarck, 1933, p. 119). He considered this to be the most important general lesson from his Moroccan studies. Next, we will look into that work in more detail, focusing on what it reveals about the social role of holiness. Westermarck regularly treats questions about holiness in the context of social conflict and competition, without explicitly pointing out that he does so.

Westermarck in Morocco

When Westermarck formulated his theoretical views on religion, he had already conducted several years' worth of field studies. His active field research in Morocco comprised altogether six years stretching from 1898 to 1913. Westermarck's first volumes on Moroccan popular religion appeared soon after the end of his first fieldwork period (1914, 1916). The First World War and other intervening events forced him subsequently to stay in Europe, but he came back in 1923, buying a house in Tangier, where he later was to stay for some months almost every year until 1939. During this later period he finished his three volumes on popular religion (RBM I-II; Westermarck, 1933) and a further book on native proverbs (1930).

Many references to Westermarck's Moroccan experiences are included in the early essays. However, in the typical manner of comparative method, they are scattered in the text as confirmations of the general argument, not as attempts to address the specific case of Moroccan religion. The later books focus on Morocco, but Westermarck still seems consciously to avoid what, in modern terms, would be called sociological analyses of the concept of the holy. He gives us examples of native ideas and practices, but he does not analyse at length what is *done* when holiness is invoked in discussion and intercourse – how the reference to 'holiness' makes a difference in a discussion[5] – nor does he explicitly describe its role in the general social dynamic of the country. Nevertheless he divulges valuable data that make such analysis possible. Against the background of our previous discussion, we now present a new reading of Westermarck's results on holiness, showing the tangible presence of contemporary Moroccan society in his work.

The holy (*baraka*) in Morocco

In Westermarck's descriptions, Morocco comes across as a place where almost anything might be religiously significant. People seemed constantly to face questions about what things or actions are holy, inauspicious or prohibited. In particular, such questions were raised when unexpected things happened: there was a wedding or a weighty decision to be made, or someone fell off a horse or a stranger came for a visit. The holy was summed up in the concept of *baraka*, or blessing. This miraculous power and gift of God took many expressions. *Baraka* was a property of certain persons – to some degree of all human beings – but also of animals, sites, foodstuffs, herbs, holidays, names, weapons and so forth. Outsiders such as idiots, madmen and strangers were strongly imbued with *baraka*. Westermarck recalls in his memoirs that his arrival to a village was seen as auspicious because it coincided with rain and the fall of food prices (MML, p. 178). People with *baraka* were honoured but also feared, as they could both cure ailments and call forth disease and death.

The worship of saints, dead and living, male and female, comes across as an important feature of the culture. It predated Islam in the region, but it had been incorporated into its monotheistic framework (RBM I, p. 51). Existing sacred sites were reinterpreted so as to associate them with specific Muslim saints. Tombs of holy men and women were agreed to have miraculous powers. They were sometimes used as asylums by criminals and insurgents at large (RBM I, pp. 559–61). Religious brotherhoods were bonds between those who shared the same patron saint.

According to Westermarck, holiness was in part hereditary in Morocco (RBM I, pp. 36–7). Prophet Mohammed had possessed the highest possible degree of *baraka*. It was passed on to his descendants via his daughter Fatima. With the spread of Islam, certain families of *shereefs*[6] partook of this inheritance. According to Westermarck, the most ancient of these *shereef* families trace their ancestry to Mûläi Idris the Elder, the founder of the royal Idrisid Dynasty (RBM I, pp. 37–8). The branch of the Wazzan Shereef was the most worshipped and influential one (RBM I, pp. 36–8). A certain tension existed between the Wazzan Shereefs and the sultans. According to popular saying, no Wazzan Shereef would ever become sultan, nor would any sultan reign without the blessing of the Wazzan Shereef. The ruling sultans also belonged to a family of *shereefs*. The sultan's *baraka* would sustain the welfare, plentiful harvests and high nativity of the country. When Sultan Hâfid took over after his brother, the deposed 'Abdl'azîz, the sardine fishery proved unusually successful. It was assumed to demonstrate the new sultan's *baraka* (RBM I, p. 39; MML, pp. 231–2).

Other families, too, had claims to *shereef* status, including the Baqqâli family,[7] whom Westermarck came to know via his close friend 'Abdsslam l-Baqqâli (Abd-es-salam El-Baqqali). But there were also 'false' *shereefs*. Moving to another region, one could fabricate a family tree and claim high connections. As a further complication, members of a family of *shereefs* could bestow their *baraka* on a client. Holy individuals could also bestow *baraka* on food and drink. They would retain their *baraka*, whereas, at other times, people might accidentally lose theirs by passing it on to others. There were also families whose *baraka* was derived from an individual saint independently of the sacred bloodline (RBM I, pp. 181–6). In principle, it was possible for any person to gain sainthood through a pious life. Heads of holy families were expected to mediate in local feuds and to lavish hospitality on visitors. The latter was possible because of a stream of donations by pilgrims at the family shrines.[8]

Politics and religion: how holiness is lost

Turn-of-the-century Morocco was an unwieldy realm. Parts of the country lay outside the sultan's reach. Meanwhile, France and Spain actively promoted their colonial interests. Two sultans ruled Morocco at the time. Mûläi[9] 'Abdl'azîz reigned from 1894. After a power struggle, his half-brother Mûläi l-Hâfid replaced him in 1908, but soon had to abdicate in favour of his brother Mûläi Yûsuf (1912–1927), who accepted European suzerainty. Morocco was reduced to a French and Spanish protectorate, although armed resistance continued throughout the colonial period. The turmoil obviously created difficulties for Westermarck. Outsiders and even royal officials were met with suspicion. Some areas (the Rif, in particular) were quite out of bounds for Europeans. But, as we will see, political conflicts were also reflected in local religious discourse. The 'sensitiveness' of the holy reflects the contested character of power relations.

Holiness could also be lost. It is of particular interest here to examine the widely circulated ideas of why the deposed Sultan 'Abdl'azîz had lost his *baraka*. According to Westermarck,

1 The court was frequented by Christians. The sultan was interested in European fashions and gadgets (RBM I, p. 229; MML, p. 215);
2 The sultan drank alcoholic beverages (RBM I, p. 238);
3 He was careless with state funds (MML, p. 215);
4 There was famine and disquiet in the country (RBM I, p. 39);
5 Perhaps the most serious of the sultan's offences: he disturbed a holy grave. A murderer had taken sanctuary at the tomb of Mûläi Idris but, contrary to custom, he was apprehended and executed. To make things worse, the murderer was a shereef and his victim a European missionary (RBM I, p. 560; MML, pp. 215–16).[10]

As we can note from the list, all the items but one (famine and disquiet) had a direct connection with the increasing involvement of foreign powers in the internal affairs of Morocco. On top of this, Westermarck recalls his own experience. When, in 1901, he visited the Dukkàla region, false rumours were spread that the sultan would let him build a home there despite local hostility (Westermarck to Tallqvist, 12 May 1901. NLF: KT), thus allegedly collaborating with a European against the locals. In particular, the sultan 'had the misfortune to get into the hands of Europeans' (MML, p. 214). Walter Harris, a *Times* correspondent who frequented the court, believed that the young sultan was benevolent but weak. When the royal court lay in Fez in 1906–1907 there was a famine, but no serious relief efforts were made. 'Abdl'azîz's rule was further obstructed by corrupt officials (Harris, [1921] 1983).

The list of the sultan's offences is illustrative of central ideas concerning *baraka*. The monarch should defend the true faith and national honour against foreigners (cf. MML, p. 231). Christians and Jews were unclean, and contact with uncleanliness was detrimental to *baraka*. Holy tombs were pilgrimage sites, sometimes out of bounds not only for Christians and Jews, but also for women and government officials (RBM I, pp. 193–6). Westermarck reports he was not allowed to visit the home of his friend Sîdi 'Abdsslam l-Baqqâli because it housed the sacred tomb of Sîdi Hosni, his grandfather. Westermarck's books and manuscript notes could, however, be kept there (RBM I, p. 65). He also wrote to a friend that he is commonly hated in a village because he has photographed its holy sites (Westermarck to Tallqvist, 12 May 1901, NLF: KT).

The sultan's crimes may appear very serious, but some individuals could break the rules without risking their inherited sainthood (RBM I, p. 238). So it was with Shereef Sîdi l-Hadj 'Abdsslam, the head of the sacred Wazzan family, who died in 1892, a few years before Westermarck's first arrival. It was said that the *shereef* was often drunk. His followers maintained that alcohol turned into honey as he drank it. When Westermarck asked why he then was intoxicated, he received the answer that everything is possible for a saint. A double miracle then, Westermarck concluded (RBM I, p. 151). This holy man married an Englishwoman (Emily Keene), but that, too, left his *baraka* untouched. Many miracles were attributed to the Wazzan Shereef. He had reportedly travelled on horseback from Tangier over the strait to Gibraltar (RBM I, p. 148). He could turn himself into a lion in order to impress his powers on Westerners who were making fun of him (RBM I, 151). Westermarck reports that the Wazzan Shereef could steal the *baraka* of his guests if any of their food was left over. Returning pilgrims would avoid his residence so as not to lose the blessing they had acquired at the holy site. One wonders why the head of the most sacred family of all Morocco needed to top up his holy powers by foul means. Furthermore, why would he not lose his *baraka* whereas the sultan would?

The fate of Sultan 'Abdl'azîz and the rumours of his loss of *baraka* indicate the close relation between *baraka* and political power. The difference between 'Abdl'azîz and the shereef of Wazzan perhaps was that the latter had popular support, whereas the former did not. The anecdote about the *shereef* turning into a lion casts him in the role of a defender of native integrity against Europeans.

The social role of *baraka*

Westermarck maintained that his researches in Morocco had shown him 'how superstition, in a quite unsuspected degree, had influenced moral ideas' (MML, p. 218). He hardly presented anything approaching a political analysis of this influence. Yet his data indicate that *baraka* was a central ingredient in conflicts of interest and competition for status. Just consider why one would fabricate a holy bloodline. M'hammed Sabour (1993), in his essay on *baraka*, simply translates *baraka* as 'symbolic capital and symbolic power' (*capital et pouvoir symbolique*). Details of Westermarck's description suggest a conflict between inherited and acquired status. The saintly families were a kind of aristocracy, but they constantly needed to reinforce their status, preferably contributing with reported miracles as proofs.[11] Inherited holiness invites jealousy, especially when the individual who has it seems to lack the required personal qualities. Hence a permanent tension prevails between inherited and acquired status.[12] The need for confirmation also applied to the sultans. One may speak of competition for legitimacy. A legitimate ruler is a successful defender of the faith and of national honour. Positions of authority must be confirmed via demonstrable connections with traditional values. In his memoirs, Westermarck analyses the sultan's shortcomings: '[t]o exercise any effective authority a sultan of Morocco must either be a saint or a warrior, but Mûlai Abd-el-Azîz was neither' (MML, p. 214).

In the typical manner of a feudal society, Morocco had a weak central government and strong local rulers. Many of Westermarck's anecdotes highlight the role of saints, especially of their graves, in a tug of war between government and local interests (RBM I, pp. 193–4, 560–1). No sultan would dare visit the sepulchre of Mûläi Brâhim in the Great Atlas. In his lifetime, this saint was expelled from Marráksh (Marrakech) because everyone, including soldiers of the Guard, would gather around him (RBM I, pp. 193–4), apparently in implicit challenge to the sultan's authority. Saints could stand up for the people of the area where they were buried, or for their own family against the sultan's. Westermarck recounts an event that supposedly took place when 'Abdl'azîz came to Dukkàla with a disciplinary expedition. Four of the great local saints appeared to him in a dream. They welcomed him, but only if he came with good intentions. 'They thereby let him know that they sided with the people, and he followed their advice' (RBM I, fn 1 p. 194). According to a modern analysis by Mohammed El-Mansour (1999, p. 56), the cult of saints in precolonial Morocco served to maintain a balance between the opposite risks of lawlessness and despotism.

The sensitiveness of the holy

Westermarck personally believed that the 'sensitiveness' of *baraka* was his most central result. Still in the spring and summer of 1939, Westermarck was telling his friends he was planning a monograph called *The Essence of Holiness* (Westermarck to Helene Arrhenius, 18 June 1939, quoted in Lagerborg, 1951, p. 446. Also see Lagerborg, 1951, p. 366). Writing to Lagerborg in the spring of 1939, he takes to task Arch-bishop Nathan Söderblom, as well as Rudolf Otto's book *Das Heilige* (and implicitly, Durkheim). These scholars 'have not considered the socially most important characteristic of the holy – its extraordinary sensitivity to external influences' (Westermarck to Lagerborg, 18 April 1939, ÅAB: RL). He is thinking of the fact that holiness could be challenged and lost, as well as transferred to new people and objects.

Westermarck was of the opinion that his results were incompatible with Durkheim's theory. The holy was not a clearly demarcated and stable category, as he believed Durkheim would have it. No systematic summary of their respective positions is offered in *Ritual and Belief in Morocco*, but Westermarck cannot resist aiming at least a blow at Durkheim. He was obviously happy with this formulation: later to reappear in *Pagan Survivals*, it reproduces an almost identical paragraph published in 1916.

> It is also sometimes impossible to distinguish between *baraka* and the profane; an instance of this is the medicinal virtue of many plants. A beneficial power is *baraka* only if it is looked upon as more or less mysterious, wonder-working, 'supernatural', not if it appears as ordinary, common, 'profane'. It is the feeling of wonder that is at the bottom of it. But this feeling allows of many degrees and may at times dwindle into insignificance or pass away; and the same is the case with *baraka*. There is not that impassable gulf between the holy and the profane which has been postulated by Durkheim.
>
> (RBM I, pp. 146–7; cf. Westermarck, 1916, pp. 75–6, 1933, p. 119)

Westermarck contrasts this with Durkheim's theory:

> Durkheim says [. . .] that in the whole history of human thought there is no other instance of two categories of things which are so profoundly different, so radically opposed to each other, as the sacred (*le sacré*) and the profane, their heterogeneity being absolute.
>
> (RBM I, fn 1 p. 147)

Westermarck's reasoning is apparently this: human feelings come in degrees and they are liable to change. Now Westermarck's results show that ascriptions of holiness similarly admit of degrees and change. This would be expected if we accept the idea that holiness is a function of feeling. An object is holy only if, and only as long as, it evokes the appropriate feelings in the perceiver. Durkheim, in contrast, thinks that ascriptions of holiness are stable. Indeed his theory implies that holiness is the single most stable category of social life, which lends stability to society as a whole.

But unlike what Westermarck believed, his data – considered simply as data – do not seem in any obvious way to conflict with Durkheim's theory if we keep in mind the difference of the main questions that the two scholars were addressing. Westermarck asked why particular objects were singled out as holy, and he gave an answer in terms of feelings. Durkheim asked where the *distinction* between the holy and the profane came from, and he gave an answer in terms of social function. Now it is imaginable that the *distinction* is a stable element of culture, while it may remain true that *ascriptions* of holiness change in individual cases. Durkheim did claim that the holy and the profane expressed a constant human tendency to view reality in terms of a fundamental dichotomy. But this is consistent with allowing for transitions from the one category to the other (Durkheim, [1912] 1968, pp. 36–9). Some other results by Westermarck also appear compatible with Durkheim's views. He reports that the unclean or dangerous is sometimes seen as holy. Murderers could cure diseases (RBM II, p. 11; Westermarck, 1933, p. 143). Conversely, saints and *shereefs* could induce illness and death. We are reminded of a Durkheimian idea of the relation between the holy and the unclean. The untouchable comes in two forms: the sacred in the usual sense (the *purely* sacred) and the unclean (*the impurely* sacred), or the sacredly auspicious and the sacredly inauspicious (Durkheim, [1912] 1968, p. 301). Westermarck, too, points to 'resemblance between *baraka* and uncleanliness'. The unclean, like the holy, 'belongs to the sphere of mystery' (RBM II, pp. 260–1; Westermarck, 1933, p. 142). This remark, indeed the phrasing, could easily be taken for *support* of Durkheimian ideas. Nor do Westermarck's data contradict Durkheim's general idea of the formative social influence of the holy. On the contrary, they demonstrate the omnipresence of *baraka* in Moroccan society. *Baraka*, among other things, served to separate the moral community from 'others' on account of their impurity.

One crucial point where Durkheim's theory seems inadequate is that holiness is, in Westermarck's material, not bestowed or removed by a centrally authorised body or 'church'. Instead, it results from competition and negotiation. The uncertainty that surrounds *baraka* may concern both social competition and the individuals' uncertainty about how to interpret a situation, an object or event. The sensitiveness of the holy implies that members of the community may have *different* ideas of what is holy. This also makes it understandable how there can be an internal, *religious* criticism of prevailing ideas of the holy – something that seems excluded by definition in Durkheim's theory. And for these reasons there may be a point in stressing the role of the individual's feelings – which will admittedly have to be considered in the context of the society where the individual lives.

Westermarck is then not simply defending Frazer's definition of religion (as something derived from awe in the face of natural phenomena) against Durkheim's. It is true that he accepts it. But his emphasis on the sensitivity of holiness highlights a dynamism that distinguishes him from both authors.[13] Like other forms of authority, holiness can be acquired and lost.

Conclusion: a dynamic view of the holy

Summing up: Westermarck believed that holiness is *sensitive*, and he believed this was because holiness depends on feeling. Durkheim, on the contrary, emphasised the *stability* of holiness. The reason was that holiness, in Durkheim's view, has a stable function in social life. It structures community members' views on reality so as to allow them to fit into a given social order. It seems to us that Westermarck's material does demonstrate the sensitivity of holiness in Moroccan society. However, this does not necessarily show that holiness depends on feeling. Judging from Westermarck's material, the reason for the sensitivity of holiness in Morocco was rather that Morocco was a *pluralistic* society. The competition for holiness was one aspect of a general competition for social capital.

By emphasising sensitiveness, Westermarck highlights the dynamic aspects of the holy. To employ an expression that Westermarck himself did not use: holiness was a matter of negotiation, not something conferred upon an individual by a homogeneous collective. We suggest that this emphasis distinguishes him from both Frazer and Durkheim. Durkheim had painted a picture of 'the community' as a unitary entity. Every aspect of religion stayed put as a support for a changeless order of things. Even in general, functionalism arguably had a built-in bias towards an emphasis on stability (see Gouldner, 1971, pp. 126–34). In contrast, Westermarck's Moroccan material offers us glimpses of a complex society where even holiness is ridden with conflict. His neglect of political analysis in a modern sense, however, did not allow him to develop this insight in full.[14]

Notes

1 In the present chapter, we will be following Westermarck's practice of treating 'the holy', 'the sacred', and 'holiness' as interchangeable terms.

2 See also Frazer (1911, p. 222): 'By religion, then, I understand a propitiation or conciliation of powers superior to man which are believed to direct and control the course of nature and human life'.

3 Cf. Frazer (1911, pp. 222–3): 'Thus defined, religion consists of two elements, a theoretical and a practical, namely, a belief in powers higher than man and an attempt to propitiate or please them. Of the two, *belief clearly comes first*, since we must believe in the existence of a divine being before we can attempt to please him' (italicisation added).

4 This leads to a problem about what to say of religious reformers who, by definition, lack the backing of a church. Rolf Lagerborg, who studied for both Westermarck and Durkheim, noted the problem and asked whether Jesus had a religion at all before he had disciples (Lagerborg, 1934, p. 116). He also points out that Jesus, to some extent, gave a new meaning to the notion of the holy (Lagerborg, 1934, p. 201).

5 However, in 'The Study of Popular Sayings' (Westermarck, 1932c, p. 208), Westermarck points out that reports of proverbs should always be given with an account of the situations where they are used.

6 *Sharif*, plural: *shurafa*; female gender: *sharifah*, plural: *sharifat*.

7 For the genealogy of the Baqqali family, see Suolinna and Lahtinen (2003).

8 A similar description has been published by Ernest Gellner (1969, p. 273), who studied Berber tribes of the Great Atlas in the 1950s. Gellner's work has been criticised (see Rosen, 2002, pp. 51–3 and, in particular, Zubaida, 2010, pp. 31–76), but this particular point about the importance of the '*sayyids*', or the bloodline of the Prophet, does not need substantiation beyond Westermarck's own analysis.

9 Mûläi = 'lord, ruler'. See RBM I, p. 36.

10 More precisely, the murderer was lured out of the sanctuary with a promise of immunity which was not honoured.

11 Cf. Maurice Bloch, describing the similar case of Yemen. Bloch, 1998, p. 149.

12 Cf. Ernest Gellner on competition for family status among the Berbers of the Great Atlas in the 1950s. Gellner, 1969, p. 273.

13 Westermarck's more dynamic take on holiness is seen at some points where Westermarck visibly disagrees with Frazer. In a commentary to Frazer's account of regicide in *The Golden Bough* (1911), Westermarck questions Frazer's claim that the new Nuer King is supposed to inherit the 'soul' of his predecessor. He argues for a different interpretation in analogy with his Moroccan data, claiming that the new king inherits his predecessor's holiness, not his soul (Westermarck, 1934b, p. 112). A soul, by definition, makes a person the person he or she is. Holiness, by contrast, is analytically distinct from the person of the individual king and may, for that reason, be conceivably acquired and lost.

14 An earlier version of this essay was presented as a workshop paper at the Nordic Network for Philosophical Anthropology conference *Questioning the Human Being* at Aarhus University, 26 August 2011. We have profited from comments by Dr Hugo Strandberg and Dr Ali Qadir.

5 Westermarck, the comparative method and the question of context

Olli Lagerspetz

Westermarck and the comparative method

The anthropological collections of the Pitt River Museum at Oxford feature a *kantele*, a Finnish folk instrument procured by Westermarck for the museum in 1893. It testifies to his friendship with the museum director, Edward Burnett Tylor (see Hirn, 1947, p. 49).[1] Tylor, then already a recognised authority on social anthropology, had helped Westermarck publish and gain visibility for his early work on human marriage, and during his visit to Oxford in 1893 Westermarck met with him as a matter of course. Tylor, appointed 1896 to the first Chair of Anthropology in the United Kingdom, was an important champion of the early paradigm of Victorian social anthropology known as *the comparative method*. Westermarck always freely acknowledged and perhaps even exaggerated his intellectual debt to Tylor, Frazer and other pioneers of the preceding generation (Shankland, 2014b).

To a modern visitor, the Pitt River Museum is a fascinating museum of a museum, its glass cases completely stuffed with objects and still largely following the organising principles introduced by its first director. It looks like something out of a Harry Potter film, a charity shop run by magicians. Westermarck's *kantele* is on display in a case labelled 'zithers'. In the same case there is a 'European' dulcimer acquired by Balfour and a 'probably English' 'bass lyre-guitar' also from Balfour; the rest are *called* 'zithers', and all have broadly oblong shapes and some strings, but come from a wide diversity of places, including Siberia. The 'Lutes' case also includes 'fiddles', that is, violins. Two cases are labelled 'Animal form in art', two are labelled 'Human form in art' and a couple of cases have baskets classified by their manufacture method or shape ('Coiled baskets', 'Plaited and twisted baskets'); further, there are glass cases for 'Methods of making fire, 'Treatment of the dead', etc.[2] The exhibits are mainly organised thematically, according to classifications by the museum staff, not geographically or by culture. The idea is to follow the development of a particular type of object across cultures.

Briefly put, the exhibition attests to the theoretical ambitions of the Comparativists. At the heart of the comparative method lay the idea that the cultural evolution of humankind could be followed by tracing the development of specific cultural traits or ideas. Cultures were classified in an evolutionary scheme from 'savage' through different levels of advancement up to modern civilisation. In reality this was only partially a principle based on chronological order, because peoples of nonindustrial societies of the *current* world were, of course, just as 'contemporary' as were the Englishmen and Frenchmen of the anthropologists' own day. The underlying idea, accepted as a matter of course, was nevertheless that nonindustrial societies represented earlier stages of cultural evolution. In his early monographs, *The History of Human Marriage* and *The Origin and Development of the Moral Ideas*, Westermarck embraced the evolutionary scheme. His ultimate aim was to demonstrate the connections of various customs or beliefs with the underlying emotions and hence with human nature through history. Among the many manifestations of moral beliefs that he considered he included marriage customs, attitudes towards homosexuality, hospitality towards strangers, ideas of punishment, ideas of the holy and many others. The development of these moral ideas was mapped from their presumable beginnings in prehuman life forms up to their present shape in civilised society. Although Westermarck in fact often emphasised the *absence* of linear progress in moral life (as opposed to the general advancement of knowledge, which he assumed to be more or less linear), he did not challenge the validity of the task itself; that is, that of isolating a particular kind of idea or custom and following it across cultures.

Westermarck on attitudes to homosexuality

To get an idea of Westermarck's method, now consider a specific example of it: his account of reactions to homosexuality in different societies. His fullest discussion of homosexuality is included in Chapter XLIII, 'Homosexual Love', of the second volume of *The Origin and Development of the Moral Ideas*, a chapter of 33 pages altogether (ODMI II, pp. 456–89). His explanation of prevalent human attitudes to homosexuality may be summed up as follows:

> [1] This censure [of homosexual practices] is no doubt, in the first place, due to that feeling of aversion or disgust which the idea of homosexual intercourse tends to call forth in normally constituted adult individuals whose sexual instincts have developed under normal conditions [. . .] whilst that particular form of it with which legislators have chiefly busied themselves evokes, in addition, a physical disgust of its own.
>
> (ODMI II, p. 483)

In other words, most individuals in the community would *personally* feel reluctant to engage in same-sex intercourse. The disgust felt by the majority at the thought of homosexual intercourse corresponds with a similar disgust felt by the sexual minority population (or 'congenital inverts') against the idea of a heterosexual relationship (ODMI II, p. 483). Second, individuals generalise their feelings of unease to cases of homosexual love by *others*:

> [2] in a society where the large majority of people are endowed with normal sexual desires their aversion to homosexuality easily develops into moral censure and finds a lasting expression in custom, law, or religious tenets.
>
> (ODMI II, p. 483–4)

Thus the instinctive reaction is in many societies institutionalised in the form of customs and laws that *prohibit* homosexual intercourse.[3] This is, however, not a universal trait of human societies:

> On the other hand, where special circumstances have given rise to widely spread homosexual practices, there will be no general feeling of disgust even in the adults, and the moral opinion of the society will be modified accordingly.
>
> (ODMI II, p. 484)

As a partly independent development, magical, religious and scientific beliefs connected with sexuality develop in human societies. Homosexual intercourse is associated with various harmful or beneficial influences. In the course of his chapter on homosexual love Westermarck cites several instances of how such intellectual elements modify the reactions of individuals:

> [3] [*Magical beliefs*:] the passive pederast may be an object of [. . .] hatred on account of his reputation for sorcery. [. . .] But the supernatural qualities or skill in magic ascribed to men who behave like women may also, instead of causing hatred, make them honoured or reverenced.
>
> (ODMI II, p. 484)

> [*Religious beliefs*:] The excessive sinfulness which was attached to homosexual love by Zoroastrianism, Hebrewism, and Christianity, had quite a special foundation. [. . .] The fact is that homosexual practices were intimately associated with the gravest of all sins: unbelief, idolatry, or heresy.
>
> (ODMI II, p. 486)

> [*Scientific beliefs*:] Very naturally, therefore, [homosexual intercourse] has come to be regarded with somewhat greater leniency by law and public opinion as they have emancipated themselves from theological doctrines. And the fresh light which the scientific study of the sexual impulse has lately thrown upon the subject of homosexuality must also necessarily influence the moral ideas relating to it, in so far as no scrutinising judge can fail to take into account the pressure which a powerful non-volitional desire exercises upon an agent's will.
>
> (ODMI II, p. 489)

In sum, Westermarck's approach assumes three levels of explanation which I have indicated by numbers in the preceding quotes. These are, first, [1] individual psychological aptitudes and thought processes, which he takes to be rooted in biological factors; second [2] social institutions that enact prohibitions and sanctions. These have an underpinning in individual emotions but also a certain independence from the latter. Finally, [3] magical, religious and other, for example, scientific, ideas enter as a third set of explanatory factors. Such beliefs belong to a social level of explanation, but they are also partly independent of law and custom, and they influence in turn *both* law and custom *and* the emotions of individuals. Each of the three sets of factors exercise influence on each of the others.

This account is in keeping with Westermarck's general view on the nature of moral judgements. It implies the objectivisation and institutionalisation of tendencies that at first only appear in the form of emotional reactions of individuals (see Fletcher, 1982, pp. 200–1; Stroup, 1982a, pp. 102–5). Westermarck's views here are, for instance, exactly analogous with his theory of the incest taboo (see Antfolk and Wolf, this volume; Segerstrale, this volume). His explanation of attitudes towards homosexuality is not simply an exercise in biological determinism, but it involves interplay between biology, social institutions and intellectual beliefs. At the same time, it must be said that he does not analyse this dynamics of interaction in detail.

Westermarck's definition of 'homosexual love' as an exercise in comparative method

Westermarck's discussion of 'homosexual love', I would like to say, follows the organising principle of The Pitt River Museum glass cases. To consider just one example, on the 29 lines that make up page 464, Westermarck relates altogether nine examples of homosexual love or related behaviour among women. He makes references to 'certain Brazilian tribes'; 'the Eastern Eskimo'; the 'Hottentot'; the 'Herero'; 'Zanzibar'; 'Egyptian harems'; 'Bali'; 'India'; and 'Greek antiquity'. This attests to the erudition with which Westermarck stunned his readers. On the other hand, these descriptions hardly amount to more than mentions, and they cannot aspire to providing satisfactory analyses of the individual cases.

The modern reader will note that the chapter on homosexual love covers not only homosexuality as understood today, but also – indeed mostly – other phenomena now described as cross-dressing, transsexuality and (male-on-male) paedophilia. Westermarck's note on Inuit women runs thus: 'there are among the Eastern Eskimo some women who refuse to accept husbands, preferring to adopt masculine manners, following the deer on the mountains, trapping and fishing for themselves' (ODMI II, p. 464). Nothing about *sexual* activity is even mentioned. As another example of (what would today be called) transsexuality, Westermarck observes that, among several peoples of Siberia, the (male) shamans would adopt female dress and behaviour:

> It happens frequently that, under the supernatural influence of one of their shamans, a Chukchi lad at sixteen years of age will suddenly relinquish his sex and imagine himself to be a woman. He adopts a woman's attire, lets his hair grow, and devotes himself altogether to female occupation. [. . .] The change of sex was usually accompanied by future shamanship; indeed, nearly all the shamans were former delinquents of their sex. [. . .] *In some cases at least* [sic] there can be no doubt that these transformations were connected with homosexual practices.
>
> (ODMI II, p. 458; italics added)

With regard to male-on-male paedophilia, it goes without saying that Westermarck discusses the widely attested sexual practices in ancient Greece. For instance, the love affair between Socrates and Alcibiades is well known from the Platonic dialogues. The important thing here is that a steady-going sexual relation between an elderly man and a boy in his teens was not considered a perversion, whereas other sexual practices in Greek society were so regarded. Another example comes from Japan:

> In the period of Japanese chivalry it was considered more heroic if a man loved a person of his own sex than if he loved a woman; and nowadays people are heard to say that in those provinces of the country where pederasty is widely spread the men are more manly and robust than in those where it does not prevail.
>
> (ODMI II, p. 477)

Westermarck was in a good position to contribute to this research because of his firsthand experience from Morocco (in his personal diary he describes 'Moors' as 'lecherous fellows'[4]). He cites the 'excessive indulgence in pederasty' among 'the mountaineers of Northern Morocco'. Westermarck also has a social explanation of this:

> Among the mountaineers of Northern Morocco the excessive indulgence in pederasty thus goes hand in hand with great isolation of the women and a very high standard of female chastity, whereas among the Arabs of the plains, who are little addicted to boy-love, the unmarried girls enjoy considerable freedom (ODMI II, p. 466–7). [. . .] So it also seems that the ignorance and dullness of Muhammedan women, which is a result of their total lack of education and their secluded life, is a cause of homosexual practices; Moors are sometimes heard to defend pederasty on the plea that the company of boys, who have always news to tell, is so much more entertaining than the company of women.
>
> (ODMI II, pp. 470–1)

Westermarck makes a similar point about the seclusion of women in ancient Greece as a contributing cause to the popularity of boys as lovers (ODMI II, pp. 470–1). A few points about his observations are worth making. First of all, in these particular cases, Westermarck does advance an explanation of cultural variation in terms of differences of social structure – namely, the seclusion vs. freedom of movement of women.[5] Although it is true that Westermarck emphasised the importance of a shared human nature at a general level, he understood the value of explaining variation. At least in some cases that he knew well, he explained variations of customs by relating them to differences in social organisation.

But second, there is some reason in all these cases to wonder about Westermarck's classification of them as instances of the *same* thing. Consider the applicability of the concept of 'pederasty' on the one hand in contemporary industrial societies and, on the other hand, in ancient Japan, Morocco and Greece. In the latter cases, love affairs between an older man and a boy were not regarded as perversions. Also, there was apparently (at least according to Westermarck's descriptions) no notion of the teenage boy being *harmed* by these practices. Similarly, considering Westermarck's examples from Siberia, one might ask to what extent the behaviour of the shaman is comparable with the modern Western practices currently described as 'transsexuality'. For instance, how are families involved in each of the cases, and what is the impact of these practices on family life?

Asking these questions implies the *disintegration* of Westermarck's relativist tenets (Cook, 1999, pp. 116–24). According to the relativist position to which he subscribed, *the same* practice – such as 'pederasty' – is condemned in some societies and viewed leniently or encouraged in others. The underlying idea must be that there is something else that is not changed and which is open to description from an objective point of view – namely, *the practice 'itself'* defined *without* reference to its role in culture. But the definition of a given practice implies answers to the question how it *is* practiced. The fact that sexuality between boys and elderly men must be practiced in secret is itself a central element of such relationships in modern Europe but not in ancient Greece. That difference has further ramifications, for instance, in family life. This means that the two practices – ancient and modern – are in important respects *not the same.*

Peter Winch on 'different' and 'the same'

In his now classical work *The Idea of a Social Science and Its Relation to Philosophy* ([1958] 1990), Peter Winch framed a main question of the book roughly in this way: What is implied in describing two instances of human behaviour as examples of the same thing? Research in both natural phenomena and social practices is dependent on the researcher's ability to make 'judgements of identity'. The researcher must be able to describe several individual things, events, etc., as being somehow instances of the same thing in order to be able to make generalising statements about the phenomenon under investigation. Speaking of two things as the same or as different implies a concept of *relevant* similarities. The question is, then, relevant to whom and for what purpose? This is, I should add, not a question of whether the two things in question are *really* the same, but for what purpose they are viewed as the same or different. 'Really the same' needs always to be cashed out as the same for this or that purpose, the same from a (specified) relevant point of view or the like.

Winch relates this question to a general comparison between the situations of a natural scientist and a student of society. The natural scientist's judgements of identity are anchored in the research tradition developed by his or her fellow scientists. The tradition allows the scientist to look for relevant similarities and differences in the material. The social scientist, in contrast, encounters not only the tradition of his or her colleagues, but also another one, namely, that of the agents whose actions he or she is studying. Insofar as regularities of behaviour are seen under the aspect of meaningfulness – which is a *sine qua non* for any study of meaningful behaviour – they are actions by individuals who make their own judgements of identity. Moreover, the agents' actions are *governed* by those judgements of identity because their ways of identifying their actions also figure in their *reasons* for performing them. Winch concludes that the relation of the social scientist to his or her objects of study is not like that of an engineer to a physical phenomenon, but more like that of an *apprentice* engineer to the actions and judgements of senior colleagues in the laboratory. Thus, for instance,

> A psychoanalyst who wished to give an account of the aetiology of neuroses amongst, say, the Trobriand Islanders, could not just apply without further reflection the concepts developed by Freud for situations arising in our own society. He would have first to investigate such things as the idea of fatherhood amongst the islanders and take into account any relevant aspects in which their idea differed from that current in his own society. And it is almost inevitable that such investigation would lead to some modification in the psychological theory appropriate for explaining neurotic behaviour in this new situation.
>
> (Winch, [1958] 1990, p. 90)

Winch makes here an implicit reference to Malinowski's *Sex and Repression in Savage Society*, where critique is voiced against Freud (Malinowski, 1927b, especially pp. 5, 85–90, 243–52).[6] Malinowski had argued that in the matrilinear societies of the Trobriands, the role of the father is less authoritarian than in contemporary European societies and, for that reason, not liable to give rise to the same kinds of conflict (Malinowski, 1927b, pp. 25–32). Later in the essay 'Understanding a Primitive Society', Winch (1972b) applied his argument about sameness to questions about magical beliefs among the Central African Azande. A crucial point in that later essay was that already the *identification* of a belief in an alien society presents the anthropologist with difficulties. The researcher needs first of all to understand what it means to profess the belief – in other words, one must get clear about the kind of role that the belief has among those who profess it. In the process, it may turn out that the belief means something quite different to the natives from what a superficially similar belief would mean in a modern Western society. Moreover, questions about the *truth* of the belief cannot be addressed unless that issue is also considered.

In *The Idea of a Social Science*, Winch illustrates his thesis concerning judgements of identity further by way of a critique of Vilfredo Pareto's distinction between 'residues' and 'derivations' in social anthropology. For Pareto, residues are recurring features of societies which will, accordingly, be suitable objects of generalisation across cultures. For instance, a large number of religions know the practice of using water or other liquids in ceremonially removing moral pollution. On the other hand, each religion has a slightly different way of motivating the efficacy of the practice. In Pareto's scheme, ideas of the latter kind are described as derivations. The derivation is simply a rationalisation of the more fundamental idea expressed in the residue (Winch, [1958] 1990, pp. 103–6). Research should concentrate on the practice in itself (the residue) and disregard the intellectual content that various cultures associate with it. Winch objects to this by pointing out that if this act of abstraction is carried out consistently, all that remains are people, or rather organisms, making certain physical movements and sounds (Winch, [1958] 1990, p. 110). We should add that even the physically salient features of the practice are likely to be lost. Movements and sounds connected with rituals of purification would hardly exhibit any *physical* unity across cultures given differences in language, rite and places of worship – by a river, in a church or temple, at home and so forth.

Abstracting away the meaningful content of the practice would strip the description of any claim to making statements about *socially* significant action. According to Winch, the social significance of an action lies in the way it is connected to a 'system of ideas or a mode of living' (Winch, [1958] 1990, p. 108). The way to investigate the practice of baptism is, consequently, to enter the ways of thinking that characterise baptism. Useful descriptions of Christian baptism would refer, for instance, to the idea of being joined to a community of Christian believers. The description should further take account of the fact that theology is not all there is to baptism; different participants are likely to give priority to different aspects of the occasion. The general point is that the question whether two practices should count as *the same* thing must in the first instance be a question directed to the individuals and institutions that perform it or live with it.

> [A]ny more reflective understanding [of a social practice] must necessarily presuppose, if it is to count as genuine understanding at all, the participant's unreflective understanding. [. . . A]lthough the reflective student of society, or of a particular mode of social life, may find it necessary to use concepts which are not taken from the forms of activity which he is investigating, [. . .] still these technical concepts of his will imply a previous understanding of those other concepts which belong to the activities under investigation.
>
> (Winch, [1958] 1990, p. 89)

Winch has sometimes been read as claiming that the social scientist must not only begin by entering the unreflective ideas of the participants but must also *stop* there.[7] As the earlier passage shows, his argument rather means that the social scientist's theoretical description must in the last analysis be something that can *connect* with the ideas of the agents. This is in a sense already inherent in the idea that meaningful behaviour is guided by motives. The individual agent is not always the best authority on what his or her motives are, but motives that we ascribe to an agent (possibly including suppressed motives) must be thought of as *intelligible to the agent* in order to count *as motives*. The language and the terms we use as researchers must be related or relatable to concepts that the agents would use. The important lesson is that when the practice in question is adequately described, no further task of causal explanation remains: to describe the ideas that go into the practice *is* to explain it – even though other questions may remain, for example, concerning historical origins.[8]

Winch's criticism is obviously applicable to the comparative method. As to functionalist sociology, its implications may depend on where the theoretical emphasis is placed. In agreement with Winch, functionalism implies that practices that prevail in a society need to be explained in terms of the *present* dynamics of the society, not, for example, merely as a residue from an earlier and now-forgotten stage of development. On the other hand, the sociological notion of function is not something that members of a society would normally employ in their social agency. Thus, functionalism conflicts with Winch's views if functions are imagined as external powers that somehow trump individual motives. However, at least the one explanation that the functionalists favoured – the need of social integration – does in the last analysis imply an oblique reference to the agents' consciousness. 'Integration' is a term for describing how the members of a society relate to each other and, thus, how they *think*. On the level of individual agency, issues of 'integration' translate to mundane concerns about, for instance, job security or family life, even though societies and the powers that influence them are far too complex for a single individual to grasp at once.

Westermarck, Winch and homosexual love

From his work elsewhere, it appears that Westermarck was not completely foreign to arguments of the kind that Winch would later use against Pareto. In a lecture in 1934 Westermarck objected to the indiscriminate use of the idea of 'survivals' in the anthropological analysis of rituals. In an argument mainly directed against Freud, he appealed to the fact that participants did not intend their ceremonies to *have* the meanings ascribed to them by Freud or by certain anthropologists (Westermarck, 2014b, pp. 173, 175). This implies also an oblique criticism of Tylor, the inventor of the concept of survival in social anthropology (cf. Warms, 2013, p. 873). We must 'be on our guard against [the] supposition, only too common in Anthropology, that similar ceremonies necessarily have their roots in similar ideas, even though practiced by different peoples' (Westermarck, 2014b, p. 175). Westermarck is not very conclusive here, but he does give a priority of sorts to explanations in terms of 'the present wants or beliefs of the people' (Westermarck, 2014b, pp. 174–5). Similarly, in his exchange with Frazer on the meaning of midsummer bonfires (RBM II, pp. 199–201),Westermarck appealed to the participants' present understanding of the rituals in question. However, Westermarck does not seem to have considered anything like Winch's fundamental point: that not only the explanation, but also the *identification*, of a practice must take account of how it is understood by the practitioners. Indeed, for Winch, the identification of the practice in a sense *is* the explanation because it is to identify the concerns that make the practice meaningful.

Westermarck's account of homosexuality is an example that Winch did not discuss, but the relevant question is now obvious. What are Westermarck's judgements of identity based on; what is his rationale for including the practices he discusses under the shared heading of 'homosexual love'? Westermarck certainly does not downplay their differences. The Siberian shaman is described as ‘effeminate’, whereas the Japanese fighter is described as ‘heroic’, ‘manly and robust’. If asked, *they* would not have thought there was much in common between them. This, of course, makes the question all the more pressing: In what sense were the Siberian shaman, the Japanese warrior and the Greek philosopher engaging in the same practice?

One answer might be that there is still something *physically* similar in each of these cases, namely, the sheer act of male-on-male sexual penetration. But even this is doubtful, considering the fact that the shaman adopts women’s clothing but does not necessarily engage in sex with other men. In any case, Westermarck is not interested in physical movements as such – which are not described *at all* in his book. His interest lies in moral ideas about the proper relations between the sexes. Another possible answer to the question of what unites Westermarck’s examples of ‘homosexual love’ might be that people in all these cases all engaged in *deviant* sexual practices. But this, too, is wrong; the descriptions plainly indicate that the practices in question were in many cases not considered deviant at all. Even when they were considered minority practices, they were not necessarily looked down upon.

The final answer seems to be that the practices at issue were deviant from the point of view of the *European* societies in which, for most of his life, Westermarck lived. Thus the leitmotif of the material is related to the social and cultural concerns of Westermarck’s own milieu. All these cases exhibit similarities of some kind with activities that, back then, were officially condemned and described without distinction as sodomy, pederasty or homosexuality.[9] The cultural situation of our time is yet different again; the classificatory framework now commonly in use does not coincide with Westermarck’s. We distinguish between homosexuality, transsexuality and paedophilia, and although sexual diversity on the whole is probably seen as more acceptable than in the Britain of Westermarck’s time, paedophilia now looms as the ultimate evil in the public imagination. In sum, the chapter on homosexual love was *Westermarck’s* way of introducing a relativising perspective on a debate that bore the stamp of *his* time and society.

Conclusions and further questions

With the advent of the geographically limited monograph as the normally expected format for anthropological research, social anthropologists grew more conscious of the specificity of the concepts in terms of which members of different cultures understood their own practices. This consciousness was stated in a general form by Winch in *The Idea of a Social Science* and in ‘Understanding a Primitive Society’. The risk of intercultural comparisons lies in *unthinkingly* classifying the comparative material, not according to how the relevant agents would see it, but how the scientist sees fit considering his or her own purposes. The lesson is not that we are not allowed to compare at all, but that we should be conscious of what we are doing. In the end it must be shown how the terms used by the researcher can be anchored back to descriptions that the agents would recognise as relevant.

Westermarck and the comparative method as a whole are hit by this criticism. The present chapter does not address the question whether the criticism would be applicable to any research carried out today; that would obviously require a new investigation. However, considering the fact that modern evolutionary approaches to human behaviour often invoke Westermarck as a predecessor, it is incumbent on them to define their position vis-à-vis the comparative method. As far as I am aware, no such general debate has yet started. But some remarks are already worth making. The search for uniformities in human behaviour constitutes the rationale for evolutionary psychology as a whole. The starting point must be that it is in principle possible to isolate a fairly specific collection of human psychological traits or behavioural tendencies that remain the same across cultures. And this invites the question on what grounds the relevant judgements of identity are made. There is certainly something prima facie problematic about routinely treating culturally very remote examples of human and, in some cases, animal behaviour as instances of the *same* thing.[10]

There are reasons to think that Westermarck's work is indeed *less* vulnerable to the critique of comparativism than modern evolutionary psychology might be. Westermarck showed an interest in cultural variation and not only in human universals. His accounts here and in other parts of his work were not based on biological determinism: he emphasised that different societies viewed 'homosexual love' in very different kinds of light. And he suggested explanations of the variation partly in terms of social structure and partly in terms of popular beliefs. As to research method, Westermarck acknowledged the importance of referring to native emotions and beliefs, not only as examples of universal human traits, but also as culturally specific explanations of particular customs.

Westermarck's fieldwork as opposed to his comparative monographs is mainly immune to the critical points raised here, and these criticisms do not diminish his historical importance as a developer of anthropological theory and practice. They do, however, call for comments from those who claim him as a predecessor.[11]

Notes

1 Westermarck's exchange of letters with Tylor has been published in Wikman (1940). In a letter, Tylor asked Westermarck to find a *kantele* (Tylor to Westermarck, 2 December 1892, ÅAB: EW, reproduced in Wikman, 1940, pp. 10–11).

2 Katherine Morris has kindly helped me with details on the exhibits.

3 Westermarck's expression, '. . . easily develops . . .' covers a development that is hardly self-evident (see the chapters by Antfolk & Wolf, Kronqvist and Pipatti, all in this volume). Elsewhere in his book he points out the transition from nonmoral reactions to moral judgements as a qualitative change that presupposes society. Westermarck, ODMI II, p. 740.

4 Westermarck, *The 'Record Note book'*, p. 3, dated 12 January 1899: 'Dessa morer äro liderliga sällar. Kanske dock, när alt går omkring, deras moral i denna punkt är bättre än vår. De göra icke anspråk på en kyskhet som de icke ega. De begå en synd, der många af oss begå två.' Available at http://filosofia.fi/se/arkiv/text/5771, accessed 26 January 2015. Transcription corrected by the present author. ('These Moors are lecherous fellows. Still perhaps, all things considered, their morals on this score are

better than ours. They do not claim chastity where they do not possess any. They commit one sin whilst many of us commit two.')

5 Later researchers have observed the further connection to differences in economy. The mountaineers of Aith Waryaghar were farmers but there was scarcity of arable land. Arranged early marriages were important for securing the availability of farmland to the family. See Suolinna (1993, pp. 45–7).

6 On p. 252 (Malinowski, 1927b), Westermarck is criticised for believing that human attitudes towards incest are to be explained by 'natural impulse' instead of family dynamics. Freud is criticised for assuming universal family dynamics based on 'innate behaviour'. At the same time, Malinowski is keen to emphasise that any family formation is a response to universal challenges inherent in biological facts connected to survival and physical maturation. See also Ihanus, this volume.

7 Cf. Kitching (2002, pp. 9–10) (italics in the original): 'Peter Winch is the Wittgensteinian philosopher who most notoriously argued that in fact social scientists and theorists should *only* use forms of description and explanation of social phenomena which are widespread or commonplace in the societies being explained, and that any notion that there could be "scientific" practices of social explanation using concepts and categories totally distinct from those commonly employed in a society rested on some fundamental "category mistake"'.

8 Historical explanation routinely makes use of ideas and knowledge that could not have been available to agents at the time. Political developments might in part be explained by climate change. Nevertheless there is something else (such as, say, crop failures) that connects with the explanation and *was* recognisable to the agents.

9 *The Concise Oxford Dictionary of Current English* still 1956 equates 'paederasty' (from Gk. *pais paidos* boy + *erastēs* lover) with 'sodomy', defined as '[c]opulation between male persons'. See pp. 855, 1195.

10 As an extreme example, consider the debate on 'rape' among insects; see Lagerspetz (2012).

11 A short version of this chapter was presented at the Nordic Evolutionary Psychology Meeting at Lammi Biological Station, Finland, 13 September 2014.

Part II

Westermarck as a precursor of evolutionary psychology

Sub-Editor: Jan Antfolk

6 Westermarck as a precursor of evolutionary psychology

The nature and nurture of evolutionary explanations

Jan Antfolk

Edward Westermarck was deeply influenced by Darwin's theory of natural selection, and this influence is apparent in several of his attempts to provide explanations of human behaviour. Westermarck went even further than most contemporary scholars and wanted to give also cultural institutions an evolutionary explanation. His most famous contribution to more recent work in sociobiology and evolutionary psychology is his explanation of why humans avoid sex with their close kin. His hypothesis that humans come to avoid sex with peers raised in the same household became known as the Westermarck effect. It would be a mistake to think that his work on incest avoidance in humans is his only contribution to sociobiology and evolutionary psychology, but the Westermarck effect is certainly the most frequently debated and well remembered. Even though it remained nothing more than a hypothesis in his days (Westermarck did not gather any data to test his hypothesis), Westermarck's hypothesis has later been used and misused, tested, supported (and also partly falsified) and discussed extensively. It continues to be at the absolute centre of discussion in the modern version of evolutionary psychology. The reasons for this are both historical and methodological. The historical reason is, I think, obvious. Alongside Darwin's work on emotions, Westermarck's work on incest was one of the first attempts to use Darwinian reasoning as a tool to understand human psychology and it thus lies at the historical roots of the entire field. It was initially greeted as the final solution to the question regarding the origins of the incest taboo, but shortly after, Westermarck's work became unfashionable and the Westermarck effect was seen as a mistaken hypothesis. Proponents of more fashionable schools of thought eagerly raised the point that this biological explanation of the origins of the incest taboo overlooked the role of culture, or they claimed that it was simply wrong in the light of empirical evidence. What Wallace and Tylor saw as a stroke of genius, Freud, Frazer, Malinowski and Durkheim saw as silly. Half a century later the Westermarck effect was nevertheless revived. In its second coming, the Westermarck effect became a paradigm case of how human behaviour could be given an evolutionary explanation. Several scholars used this particular explanation as a weapon in the battle between proponents of biological vs. cultural influence in the nature vs. nurture debate. Initially it was seen as a solid and make-sense 'biologism' to be used in favour of 'nature'. Even some scholars who might have considered themselves sceptical about evolutionary explanations of human behaviour seem to have agreed that the Westermarck effect indicated that biology was not completely unimportant. But some 'naturalists' also saw Westermarck's idea as an intriguing example of the interaction between nature and nurture. In my opinion, it continues to stand as one of the most elegantly formulated examples of how and why no clear and meaningful distinction between nature and nurture can be drawn. Westermarck showed that an evolutionary effect could rest on what is indeed a social condition. Proposing that coresidence in childhood is part of how we come to avoid inbreeding, he showed that, evolutionarily speaking, both culture and nature constitute biological constraints and yield evolutionary possibilities. Westermarck's argument also contains several of the important aspects one needs to consider when giving a phenomenon an evolutionary explanation.

When we look for explanations, we tend to look for a story of causation. Causation is part of what distinguishes explanation from description. A description states that 'X is doing Y', and an explanation states that 'X is doing Y because of Z'. Causal explanations come in many forms, and when we are concerned with evolutionary explanations we need to make a distinction between *ultimate* and *proximate* causation. Although this distinction is not unique to evolutionary thought and has been made in some form by philosophers of almost all traditions, ranging from Aristotle to Heidegger, it is very important to Darwinists. The ultimate causation of any adaptation always lies in natural selection and the mechanisms with which it acts. The force of selection pressure and the room for adaptive change can vary, but unless the discussed phenomenon or phenotype has yielded a reproductive advantage over its alternatives, it cannot be one that has been selected for in the now relevant sense by natural (or cultural) conditions. Thereby it is not an adaptation in the Darwinian meaning of the word (although it can be a by-product of an adaptation). Thus, a description (with or without factual support) of the presumed reproductive advantage of the feature we want to explain is key to any evolutionary explanation. This explanation purports to give a historical account of why a certain feature exists and why it is more common than its alternatives. On the other hand, *proximate* causation explains the mechanisms involved in bringing about this feature in the physiological or behavioural repertoire of a particular individual. It involves both developmental and 'mechanistic' aspects. Focusing on behaviour, developmental aspects might involve, for example, learning and past experience. What I here call 'mechanistic' aspects might involve the physiological changes in the body that bring about a certain feature in a certain situation. An example of these 'mechanistic' aspects is the hormonal changes that precede experiencing fear and taking a step back when one sees a snake on the ground. A full evolutionary account of our typical reaction to dangerous animals would then look for ultimate causation (for instance, pointing out that avoiding snakes gives the agent a reproductive advantage over approaching them because the survival benefit of approaching a snake [there are better and safer food sources] is outweighed by the possibility of a detrimental bite) and proximate causes (e.g., the preparedness of the nervous system to modulate its responses to snakes and snakelike objects based on past experience). Logically speaking, ultimate causation does not dictate any particular shape of proximate machinery, but unless there is *some* proximate machinery that can be phylogenetically shaped by reproductive variation related to this feature, there is no evolutionary effect to be found.

The Westermarck effect indicates an intriguing proximate mechanism that modulates our sexual behaviour. In this case, growing up in propinquity is the social hinge upon which human incest avoidance rests. Living conditions (at least when discussing humans) are often considered part of a cultural or social domain and much less part of biology. But Westermarck believed that because growing up together is often a natural consequence of having a sibling, and inbreeding is, in most cases, a reproductive drawback, then the reproductive advantage of avoiding inbreeding can ultimately cause the appropriate proximate machinery to arise and inhibit sexual attraction between individuals who grow up together.

As noted earlier, the distinction between proximate and ultimate causation is important to evolutionary psychologists. It is a distinction that helps separate 'how' from 'why' and a distinction seen as more meaningful than the rather obscure distinction between culture and biology or between nature and nurture; and, as we will see in the following chapters, the distinction must be made in order to give a meaningful evolutionary explanation to a given form of behaviour.

I earlier separated between descriptions and explanations, saying that only the latter provide a story of causation. But it is important to remember that particular forms of behaviour (or lack thereof) can be explained in several ways. Evolutionary theory provides a certain *type* of explanation. Much like any other type of explanation, evolutionary explanations are based on certain assumptions, and they give certain types of answers. To see which these assumptions are and which methodological constraints evolutionary theory has, they can be contrasted with other types of explanation. Here, the discussion will be limited to two very general groups of explanations: intentional and evolutionary explanations. By contrasting these two, we might be in a better position to understand the role of evolutionary explanations. Moreover, we might also better see when the latter are meaningful and when they are not.

Consider, for example, a child eating an ice cream on a sunny summer day. I guess that in most cases we would not feel it necessary to ask for a reason, as we would not be very surprised at seeing this particular behaviour. If we were, however, to ask for an explanation, we would likely be trying to understand what led this particular child to eat ice cream at this point in time. We can certainly imagine several types of questions to ask, all of them looking for slightly different types of explanation. Perhaps we ask the child: 'Why are you having ice cream?' This is a common type of question, indeed part of our social nature, of our seemingly inherent wish to understand the motives of other people, or the need felt by agents themselves to justify something that, to us, appears as slightly odd. If we were to ask the child this question, I assume the child might answer that he or she *likes* ice cream, especially when it's sunny and warm – and that's why.[1] I assume this is a correct answer (at least it would be absurd to think the child is 'wrong') because it satisfies our wish to know something concerning *the child's* reasons for what he or she is doing. In this case, the child's behaviour is given an *intentional* explanation: the child has intentionally taken the necessary steps to satisfy a particular desire. Perhaps the child has asked its parents for permission to have an ice cream, walked over to a nearby stand and then traded a coin for a cone. In this particular case, this kind of explanation would be satisfactory and we would accept the reported reason as an answer to our question. I think I can safely assume we would feel quite unsatisfied if the child said it had no reason at all. Perhaps we would even think that there must be a reason and that the child is either unaware of this reason or that he or she refuses to tell us. On the other hand, if we ask another child why he or she is *not* having ice cream, we might, of course, get a slightly different response. Perhaps this other child also desires ice cream (and seeing other children eat ice cream, this desire might even be stronger) but does not have the means to satisfy his or her desire. Perhaps the parents did not give their permission, or perhaps the child has lost the money. We could also think of this as a reasonable explanation. Another very reasonable explanation could be that the child does not like ice cream or wants something else instead.

The characteristic feature of explanations of this type is that they allow us to 'understand' someone in terms of their intentions or lack thereof.[2] They seem to do a good job when we want to understand other people and their behaviour in day-to-day interactions. We tend to view other humans (and at least some other animal species) as intentional beings – beings that experience desires that they wish to satisfy and act according to what they perceive as desirable and appropriate rather than undesirable and inappropriate. We would, on the other hand, be surprised to hear that a robot eats ice cream because it *likes* it or that a car does not start because it lacks the desire to do so. But when we are concerned with other humans (and I think this is more true for other humans than it is for individuals belonging to other species) we recognise that they have desires and, equally important, that there is variation in the desires they do have. One likes ice cream, another does not. One likes lemonade, another likes both. But we understand what a desire is and that it calls for being satisfied. And because of this, intentional explanations are meaningful.

We can also be interested in other types of explanation. We might, for example, feel that the type of intentional explanation given in the previous example is not one of high resolution enough, or even that a particular agent's intentions are irrelevant or insufficient for answering our question.[3] Then we might pose follow-up questions regarding what brought this child to like ice cream in the first place (we can imagine asking the child regarding this aspect, but we cannot be certain about whether or not the child will be aware of the correct answer), or we might be curious about how this desire is manifest on a neurophysiological level (we would likely not ask the child regarding this). In our curiosity we could uncover a long chain of physiological cogwheels in a machine that involves various components. Here we assume that one event is causing another and all of them lead up to the behaviour we aim to explain. And as the number of cogwheels increases, we describe things with increasingly improved resolution. We might learn to know the neural correlates of certain desires, the metabolic events that invariably precede ice cream consumption desires and the biochemical processes involved in satisfying the desire for ice cream by ingesting ice cream molecules.[4] Bit by bit we uncover, on a mechanistic level, the *proximate* causation of the behaviour we want to explain.

This, I think, is the type of question that traditionally has occupied psychology. Irrespective of whether behaviour has been explained in terms of bodily processes or learning, or explained in terms of some new and impressive psychological vocabulary, they have generally constituted what can be considered proximate explanations. Evolutionary psychology, since the days of Darwin and Westermarck, has also been concerned with answering questions regarding the *ultimate* causation of a range of human behaviours. Today the range of behaviours given an evolutionary explanation has expanded (and likely far beyond what is meaningful), but Westermarck's explanation of why we avoid incest was one of the first such attempts. This was a question where intentional and evolutionary explanations might arrive at wildly different answers. As a psychologist with an interest in why people avoid incest (and why, at times, they do not), I tend to ask people why *they* do not have sex with their kin. Most people find the question nonsensical and absurd. This seems to be a case of what William James considered a problem to see beyond our instincts (James, 1890), something either so normal it cannot be thought of as strange or something so far outside the usual repertoire of human behaviour that people lack conscious cognition regarding it. This makes it hard to provide intentional explanations in this case. But when someone does give an intentional explanation here, they usually say that sex with close kin would be disgusting – and that's why! The evolutionary explanation wants to go a few steps further and ask exactly why incest is disgusting or why our family members so seldom appear as possible sexual partners in our minds.

Twelve decades after Westermarck provided his explanation how and why we avoid sex with our family members, we have learned more regarding the processes involved. Building directly on his work, psychologists, anthropologists and biologists find ways to learn more about the interaction between culture and biology that is part of human incest avoidance. In this endeavour Westermarck's legacy is apparent. His work continues to yield testable research questions and inspire debate.

Notes

1 It is important to note that questions are contextual, and what is considered an acceptable answer depends on our own motives to pose a certain question. Although the response 'I *like* ice cream' could constitute an acceptable answer to the question, 'Why are you having ice cream?' in some cases, this response would seem quite impertinent if our reason for asking was that we were surprised to see that the child was eating ice cream instead of having a decent meal for lunch.

2 This is indeed a very particular sort of intentional explanation, in which 'intentional' is limited to a mental or psychological state (desire) of the actor. Other possible types of explanation can also be considered intentional.

3 Imagine, for example, that we are interested in why people generally consume ice cream and how this sociocultural phenomenon has come to be. Then we would be asking for a historical explanation and likely direct our question at someone who is considered an expert on the topic.

4 We can imagine several instances where we eat ice cream without a particular desire for ice cream. Even if we do not like ice cream, we might resort to having ice cream if there is nothing else to have and we feel very hungry.

7 Itemising Westermarck's hypothesis

The assumptions embedded in Westermarck's explanation of human incest avoidance

Jan Antfolk and Arthur P. Wolf

One of Westermarck's most debated contributions was his hypothesis that natural selection has shaped human sexuality to not encompass sexual attraction to individuals with whom one coresided in early youth. Because such individuals tend to be close kin, Westermarck argued that this aversion appears as a sexual aversion between close kin. The debate surrounding Westermarck's explanation of incest avoidance has often been imprecise, and his explanation has often been treated as an uncomplicated and simple hypothesis. We argue, however, that it contains at least four assumptions: 1) *The assumption of ultimate evolutionary causation;* 2) *The assumption of childhood coresidence as the proximate individual-level activation;* 3) *The assumption of imperfect functioning;* and 4) *The assumption of a social extension of the individual aversion.* Although all these assumptions form parts of a coherent argument, they are as such logically separate. This implies that supporting or falsifying one of these assumptions does not necessarily lead to a support or falsification of Westermarck's explanation of incest avoidance in its entirety. By itemising Westermarck's explanation, the limitations of the philosophical arguments and empirical evidence aimed at supporting or falsifying Westermarck's explanation can be better understood. The aim of the present chapter is thus first to introduce Westermarck's hypothesis regarding the natural origins of the human inbreeding avoidance and the incest taboo. Second, we will itemise the hypothesis, arguing that, rather than being one simple hypothesis, it contains at least four independent assumptions. Third, each of the assumptions embedded in Westermarck's reasoning will be discussed in the light of philosophical arguments and empirical data that have been presented as supporting or falsifying Westermarck's position. The aim of the chapter is not, however, to give a full historical account of the debate surrounding Westermarck's reasoning (see, e.g., Wolf, 1995; Wolf & Durham, 2005 for a more complete historical account). Rather, we wish to bring forth the limitations of the philosophical arguments and empirical evidence aimed at supporting or falsifying Westermarck's explanation.

Westermarck's hypothesis regarding the natural origins of the human incest avoidance and the incest taboo

One of Edward Westermarck's major scientific contributions – and likely his most extensively debated one – is his hypothesis regarding the development of an apparent aversion to sexual relationships between family members. In his book HHM, Westermarck (1891a) proposed that sexual aversion develops between two individuals who live in close propinquity during childhood. Because such individuals tend to be genetically related, the sexual aversion manifests itself as an aversion to sexual relationships between close kin. Embedding his hypothesis in an evolutionary framework, popular in the late 19th century, his hypothesis rapidly gained great recognition amid the scholarly elite of his day. However, along with the rise of Freudian psychology and modern sociology, Westermarck's hypothesis lost its fame just a few decades later (Wolf & Durham, 2005). Since the late 20th century, with the renewed interest in explaining human behaviour in evolutionary terms, Westermarck's hypothesis has once again received attention. It can be argued that during these three different phases, unfortunately little attention has been paid to Westermarck's actual reasoning. This has, in turn, led to unwarranted simplifications and, consequently, a rather confused debate. The discussion has often contained misinformed counterarguments rather than solid logical arguments or empirical evidence that actually pertains to Westermarck's hypothesis in its entirety.

In HHM, Westermarck famously proposed that human incest aversion (i.e., the apparent aversion to sexual relationships between related individuals) is a natural aversion activated between two individuals who live in close propinquity during childhood.

> What I maintain is, that there is an innate aversion to sexual intercourse between persons living very closely together from early youth, and that, as such persons are in most cases related, this feeling displays itself as a horror of intercourse between near kin.
>
> (HHM, p. 320)

Westermarck argued that during the course of evolution, human incest aversion has developed through natural selection. Perhaps more lucidly than anywhere else in his work, he explained his position in 1889, in a speech directed to a student society (Nyländska Afdelningen, an association for students hailing from the province of Nyland in Finland). Summing up his argument, Westermarck stated:

> And it is probable that this instinct developed through natural selection. We cannot give away to carnal love with those with whom since our earliest childhood we have lived in propinquity, to whom we generally stand as siblings, parents, or children; since each such passion must have gone extinct in our species' childhood through the increased mortality to which offspring from closely related parents are subjected.
>
> (Westermarck, 2014a, p. 155)

He suggested that the ultimate selective force driving human sexual inclinations away from inbreeding is the biological costs of inbreeding.

> That [marriages between close relatives] really bring about defects in the offspring has admittedly been disputed, but, as it appears to me, on false

> grounds. Several experiments conducted on animals, as well as several facts drawn from the field of anthropology, seem to demonstrate that offspring from very closely related parents are relatively weaker than others.
>
> (Westermarck, 2014a, p. 155)

Westermarck also assumed that this individual aversion gave rise to the social taboo that prohibits incest in most known cultures. The reasoning on this point was, however, somewhat less transparent.

> Marriage between close kin is to us something immoral. Why? Because there is a given innate aversion toward sexual relationships between persons that since early childhood have lived together on a daily basis; in consequence of such liaisons against our natural instincts.
>
> (Westermarck, 2014a, p 155)

Although the exact process by which individual emotions are turned into general norms is not clearly explicated in Westermarck's work (see Pipatti, Chapter 10, and Kronqvist, Chapter 12, both in the present volume, for Westermarck's sentimentalist approach to human morality), he, nonetheless, argues that the incest taboo is a social extension of the individual aversion.

The four assumptions embedded in the Westermarck hypothesis

Although Westermarck's reasoning often has been taken as a uniform hypothesis (e.g., Freud, [1912–13] 1946; van den Berghe, 1980; Williams & Finkelhor, 1995), it seems to contain a number of assumptions logically separate from one another. Westermarck's explanation of human incest avoidance, as presented earlier, entails at least the following four crucial assumptions:

1. *The assumption of ultimate evolutionary causation.* As the ultimate evolutionary cause Westermarck proposed the 'the increased mortality to which offspring from closely related parents are subjected' to be the selective force modifying human sexual psychology to exclude, in general, sexual attraction to close kin.
2. *The assumption of childhood coresidence as the proximate individual-level activation.* As the proximate cause of sexual aversion on the individual level, Westermarck proposed that 'living very closely together from early youth' causes the activation of an innate disposition to develop a sexual aversion between two individuals.
3. *The assumption of imperfect functioning.* Westermarck further assumed that this proximate activation is, evolutionarily speaking, imperfect and does not always lead to a development of sexual aversion between *close kin.* Rather, as he suggested, persons living together in early youth are 'in most cases related, this feeling displays itself as a horror of intercourse between near kin'.

4 *The assumption of a social extension of the individual aversion.* Westermarck argued that the aversions felt on an individual level are generalised into the social taboo prohibiting incestuous acts.

Importantly, these four assumptions embedded in Westermarck's reasoning are logically separate. The assumption of ultimate evolutionary causation is the only assumption that is clearly adaptationist, suggesting that the general avoidance of inbreeding is the result of natural selection. However, this assumption in itself does not necessitate that the proximate cause of activation at the individual level is proximity in early youth. It is imaginable that living together in early youth may activate a sexual aversion, regardless of whether this process has an evolutionary underpinning or not. For example, it is likely that growing up with another individual causes several types of emotions and also a will to exist in and maintain the nature of the family relationship. This, in turn, may cause an individual to adopt the norms that regulate family relations. Similarly, the assumption that the process regulating incestuous inclinations is imperfect (i.e., unable to discriminate between nonkin and kin) is an assumption that does not necessitate either the ultimate evolutionary cause or the type of proximal activation suggested. Any mechanism – naturally selected or otherwise – can be imperfect. Hence, this assumption neither necessitates nor dismisses evolutionary causation. Finally, the explanation of social incest taboo as an extension of generalised individual aversions does not depend on the aversion being an adaptation, on proximity in early youth being the proximal activation, or on this proximal activation being unable to separate between nonkin and kin.

The fact that the aforementioned assumptions are logically separate does not mean that they do not constitute parts of a stringent and reasonable argument. It has, however, another important implication. The implication is that falsifying one assumption does not lead to a falsification of the other assumptions, and analogously, supporting one assumption does not lead to a support of the other assumptions.

Philosophical and empirical attempts to support and falsify Westermarck's explanation

Westermarck's explanation quickly gained fame among his contemporary academics. Roughly two decades after the original presentation of his explanation, a number of scholars nevertheless raised vehement philosophical criticism against Westermarck. This criticism finally led Westermarck's hypothesis to lose popularity. Being almost completely forgotten for a large part of the first half of the 20th century, a renewed interest in Westermarck's explanation was brought about as a consequence of the rise of sociobiology in the 1960s (Talmon, 1964; Wolf, 1970) and evolutionary psychology in the 1980s (e.g., Parker & Parker, 1986; van den Berghe, 1980). Since Westermarck presented his explanation of human incest avoidance in 1891, theoretical and philosophical arguments, as well as empirical evidence, have been gathered in attempts to support or falsify Westermarck's explanation. Although some effort has been put into addressing whether Westermarck was right or wrong, there seems to be no consensus on whether his explanation of the origin of the human inbreeding aversion is correct or incorrect. There are at least two reasons for this: First, Westermarck may have been correct on one of his four assumptions and wrong on any of the other. Second, each of these assumptions is notoriously difficult to test empirically. Only the second assumption, the assumption of childhood coresidence as the proximate individual-level cause of aversion, and the third assumption, the assumption of imperfect functioning, could possibly be tested with true experiments, but such experiments cannot be carried out on humans for obvious ethical reasons. However, some empirical studies indirectly pertain to each of these assumptions, and each assumption can be put under theoretical and philosophical analysis. Importantly, by itemising Westermarck's explanation, we are more likely to increase the precision in the debate.

1) The assumption of ultimate evolutionary causation

Although more precise knowledge about the costs of inbreeding may have been unavailable at the time when Westermarck expounded his argument, more recent evidence supports the notion that 'offspring from very closely related parents are relatively weaker than others'. Offspring from sexual relations between first-degree relatives have been shown to be markedly less fit (i.e., more likely to have diseases, as well as having higher mortality rates and lower reproduction) than children from nonincestuous sexual relations (Adams & Neel, 1967; Aoki, 2004; Bittles, 1983; May, 1979; Seemanová, 1971). This fitness decrease has been explained as a consequence of the increased allelic homogeneity due to the increased likelihood of inheriting the same alleles from two related parents. Another overlapping explanation is that inbreeding increases the likelihood of harmful expressions of recessive alleles (Bittles, 1983; May, 1979). The known distinct decrease in reproductive fitness suggests a strong selective pressure against inbreeding in humans.

That inbreeding evidently decreases fitness provides a plausible and parsimonious evolutionary explanation of why humans avoid sexual relationships with close relatives. However, the phylogenetic argument is perhaps even more compelling. It has been shown that in many species (e.g., Paul & Kuester, 2004; Pusey & Wolf, 1996), inbreeding is avoided in natural circumstances. Still, the biological cost of inbreeding neither provides absolute and compelling evidence for the hypothesis that humans avoid having sexual relationships with close kin ultimately due to an evolutionary cause, nor does it imply that there is *no other* cause. However, to this date, no other theory has been put forward that parsimoniously explains the existence of inbreeding avoidance over a number of species and human cultures.

2) The assumption of childhood coresidence as the proximate individual-level activation

The assumption of childhood coresidence as the social hinge upon which the individual-level activation hangs has perhaps been the most debated part of

Westermarck's explanation. Rather than a purely biological explanation, the proximate explanation proposed by Westermarck is an environmental and social activation of the aversion at the individual level, that is, coresidence in childhood. He argued that the aversion is activated 'between persons living very closely together from early youth' – a social cue which must have been a valid indicator of actual relatedness in order to function as the proximate mechanism of inbreeding aversion in humans.

Although the proximate explanation embedded in Westermarck's explanation was empirically unsupported, it rapidly gained recognition amid the scholarly elite of his day. In fact, for most scholars, this assumption in itself constitutes 'the Westermarck hypothesis'. For example, Edward Burnett Tylor wrote to Westermarck, saying that 'I should not be surprised to find [your hypothesis] or something like it the real solution'(Tylor to Westermarck, 23 October 1892, ÅAB: EW, published in Wikman, 1940, p. 9). In the foreword to HHM, Alfred Wallace noted the book's 'ingenious and philosophical explanation of the repugnance to marriage between near relatives' (HHM, p. vi) and wrote to Westermarck saying, 'I like your discussion of the origin of universal repugnance to incest very much, and I think you have solved the problem' (Wallace to Westermarck, 13 April 1891, ÅAB: EW, published in Wikman, 1940, p. 18). However, Westermarck provided neither direct nor indirect support of this assumption, and the scientific community seems not to have required one. His explanation was accepted without being empirically tested. As Westermarck's hypothesis found fame again in the late 20th century, some scholars have tried to find ways to test this particular assumption.

The first piece of evidence drawn to support for this assumption stems from anthropological research. Anthropological data has been gathered mainly (but not exclusively) from two different cultural settings: the *Sim-Pua* marriages in Taiwan (Wolf, 1970; Wolf & Huang, 1980) and children cohabiting in Israeli *kibbutzim* (Lieberman & Lobel, 2011; Shepher, 1971; Spiro, 1958; Talmon, 1964). These settings have been taken to provide natural experiments of the effect of childhood propinquity on unrelated individuals. In 19th-century Taiwan, young girls were frequently adopted by their later parents-in-law to be raised together with their future husbands. In these marriages, a girl was adopted at a young age to a family with a young boy, who later became her husband. Examining these marriages from demographical records with a total sample of 15,000 couples, it was found that in these marriages the fertility was lowered by 30% and that divorce rates relative to other comparable arranged marriages were increased (Wolf, 1970; Wolf & Huang, 1980). After controlling for several factors, such as health differences between the adopted and nonadopted brides, Wolf (1970, 1995) concluded that only the spouses' shared childhood residence could explain this difference, that is, the decreased sexual interest between the partners.

Talmon (1964) studied adult couples of individuals raised in Israeli *kibbutzim*, where nonrelated children are raised together. She found that none of the partners had been raised in the same peer group and that in only four cases the partners were raised in the same *kibbutz* but in different peer groups, suggesting that childhood propinquity decreased the likelihood of entering marriage. Shepher (1971) reported that children raised at the *kibbutzim* spent most of the day together: they bathed, ate and played together. Hence, there was an abundance of physical and emotional contact between these children, and at times boys and girls slept in the same bed. Of the 65 studied adolescents and adults, there were no reported sexual affairs or marriages between members of the same peer group.

In more recent years, Bevc and Silverman (1993) have investigated how propinquity between opposite-sex siblings influenced the likelihood of engaging in incestuous behaviour. Individuals reporting incestuous behaviour were compared to a matched control group reporting no incestuous behaviour. A significantly higher prevalence of incestuous behaviour, including penetrative sexual activity, was reported amongst siblings separated during childhood for more than one year. Bevc and Silverman (2000) replicated their study with a new sample. Based on reported sexual behaviour, respondents were categorised into three different groups: genital intercourse, other sexual activities and no sexual activities. The results from this study were coherent with earlier results: 31.5% of the siblings reporting genital intercourse had been separated more than one year before age 10, whereas this was true for only 2.9% of the siblings reporting other sexual activities and 3.8% reporting no sexual activities, respectively. Further analyses showed that 76% of the sibling pairs reporting intercourse had not lived together at all during infancy (Bevc & Silverman, 2000).

Taken together, these studies indicate that coresidence during childhood decreases sexual interest between two individuals, as measured by the proxy variables employed in these studies. Although the anthropological studies show that coresidence in childhood is enough to decrease the sexual attraction in adulthood, there is probably more to human inbreeding avoidance than coresidence in childhood. On the one hand, the studies do not show a *direct* causal link between coresidence and inbreeding aversion. On the other hand, several evolutionarily minded scholars would argue that Westermarck was only partly correct. In fact, coresidence seems to be one cue of biological relatedness to another individual, but it is unlikely the only cue. For siblings it seems that coresidence is only a secondary cue. A more primary cue of relatedness is seeing a mother take care of a younger infant (Lieberman, Tooby, & Cosmides, 2007). Only in younger siblings, who do not have the possibility to see their mother breastfeeding and nursing an older sibling, is coresidence used to regulate sexual aversions towards the other individual. For other types of family dyads, for example, parent–child dyads, other cues are used (e.g., Antfolk et al., 2014; Lieberman, Antfolk, & Santtila, submitted; Tal & Lieberman, 2007).

3) The assumption of imperfect functioning

Westermarck suggested that the mechanism activating the aversion on an individual level is coresidence in childhood. As this occurs mostly with related individuals, the aversion is generally, but not exclusively, felt between close kin. Even if the detrimental genetic costs in the offspring only occur if the parents are biologically related, the proximate activation of the aversion towards entering into sexual relations with persons with whom one coresided in early childhood does not, as such, differentiate between biological and nonbiological dyads. Thus, Westermarck's hypothesis allows for interpreting avoidance of sexual activity between nonrelated family members as an evolutionary by-product. In fact, avoiding sexual relationships with a nonbiological sibling is biologically costly, as this may result in a lost opportunity to gain offspring with no risk of increased fitness costs (Lieberman & Antfolk, 2015). Such a mechanism can be selected for only if the benefits of avoiding sex with relatives outweigh the costs of avoiding sex with coresident nonrelatives.

Studies on anthropological data from Taiwan and the Middle East are also pertinent to the assumption of imperfect functioning. In all the aforementioned studies, the samples employed were not biological siblings. Rather, the sample consisted of unrelated children raised together under circumstances, which at least to some extent, were approved of by the community in which it occurred. As such, these studies provide an interesting empirical test of the assumption that the proximate mechanism that regulates inbreeding aversion in humans is imperfect. As Westermarck suggested, this proximate mechanism can take the role of regulating inbreeding aversion only if the individuals coresiding in childhood in *most* cases are related, but it is not required that they should *always* be related. Quite importantly, any empirical test that investigates the effects of coresidence would risk confounding other aspects, such as increased familial emotions between two individuals and receiving care and investment from the same adults with coresidence per se. To this extent, the anthropological data is the best available data to support the assumption that the proximate mechanism functions imperfectly.

Thus, some of the aforementioned studies can be taken as providing support to both the second assumption, where childhood coresidence is seen as the proximate individual-level activation, and the third assumption, that this proximate activation does not distinguish between kin and nonkin. Yet, it is important to note that these studies are not true experiments. Furthermore, even if much effort has been put into controlling for other factors that may have been different between individuals raised together and individuals not raised together, it is impossible to rule out the differences between the comparison groups. For example, different traditions inevitably involve different cultural expectations to some extent. Westermarck claimed that coresidence in childhood decreased sexual attraction in adulthood, but he did not say that this occurred miraculously. Indeed, any route such a phenomenon could take – cultural or otherwise – a decrease in sexual attraction ultimately caused by childhood coresidence would support his assumption. However, only to the extent that these 'natural experiments' can be considered methodologically sound, can the data these studies provide be interpreted as support for Westermarck's hypothesis.

4) The assumption of social extension of individual aversion

Perhaps the most muddled assumption embedded in Westermarck's hypothesis regarding inbreeding avoidance is his argument that the incest taboo, which is present in most human cultures, originated from individually felt aversions. Whereas many considered this explanation to be reasonable, some scholars felt that Westermarck's reasoning was inherently counterintuitive. For example, Sir James Frazer (1910, p. 97) wrote, 'law only forbids men to do what their instincts incline them to do; what nature prohibits and punishes, it would be superfluous for the law to prohibit and punish'. This type of counterargument was then repeatedly directed towards Westermarck's reasoning by, amongst others, Sigmund Freud and Claude Lévi-Strauss. Both argued that the prohibition Westermarck was trying to explain, in fact, disproved the explanation he offered. As Wolf (e.g., 2011) has pointed out, Freud has often been quoted by other academics as successfully falsifying Westermarck's hypothesis. In his book *Totem and Taboo*, Freud ([1912–13] 1946) delivered an even more consequential, albeit empirically unfounded, criticism of Westermarck's hypothesis. Freud quoted Frazer's criticism and then added,

> To this valuable argument of Frazer's I can add that the experiences of psychoanalysis make the assumption of such an innate aversion to incestuous relations altogether impossible. They have taught, on the contrary, that the first sexual impulses of the young are regularly of an incestuous nature and that such repressed impulses play a rôle which can hardly be overestimated as the motive power of later neuroses.
>
> (Freud, [1912–13] 1946, p. 160)

What Freud meant with 'the experiences of psychoanalysis' is unclear. Psychoanalysis did not empirically *show* that the relationships within a family were sexual. Psychoanalysis offered little more than speculation, warranted only by introspective contemplation and a few case studies of abnormally functioning individuals. Subscribing to Von Baer's and Haeckel's theory of ontogeny (Gould, 1977), Freud inaccurately assumed that the cultural history of man was repeated in the psychological development of each individual ([1912–13] 1946). Theoretically confusing and empirically unsupported, Freud's explanation still became popular. For example, referring to Freud, Lévi-Strauss (1969, p. 17) argued that 'psychoanalysis finds a universal phenomenon not in the repugnance towards incestuous relationships, but on the contrary, in the pursuit of such relationships', and Murdock (1949, p. 291) asserted that Westermarck's hypothesis of human incest avoidance 'flagrantly overlooks, and even inverts, that vast body of clinical evidence which shows that incestuous desires are regularly engendered within the nuclear family and are kept in restraint only through persistent social pressure and individual repression'. Considering psychoanalysis and the clinical setting as a source for evidence, these refutations implied that incestuous inclinations are common, so common that Freud wrote ([1912–13] 1946, p. 24), 'Psychoanalysis has taught us that the first object selection of the boy is of an incestuous nature and that it is directed to the forbidden objects, the mother and the sister'. Hence, the apparent lack of incestuous sexual activity must be due to the universal but cultural prohibition against incest (see Ihanus, Chapter 3, this volume). The Freudian position builds on the assumption that incestuous desires are common. However, Westermarck had much earlier already pointed out that:

> The law can prohibit a brother from marrying his sister, but it cannot prohibit him from being ignited by carnal love of his sister. Yet, where is this love? It is not laws, nor customs, nor nurture that keeps the home pure of criminal liaisons, but an *instinct*, which under normal circumstances makes carnal love between close relatives a psychological impossibility.
>
> (Westermarck, 2014a, p. 155)

Almost as if his argument was directed against Freud, Westermarck emphasised that if humans avoided incest due to a social taboo prohibiting sexual relationships between close kin, we would still find people to feel a sexual desire towards their kin. 'Yet, where is this love?' Westermarck presented a sound analogical argument for his position: a social taboo is unlikely to extinct an existing desire. Much like the social norms proscribing and even forbidding homosexuality, '[t]he sexual instinct can hardly be changed by prescriptions; I doubt whether all laws against homosexual intercourse, even the most draconic, have ever been able to extinguish the peculiar desire of anybody born with homosexual tendencies' (HHM II, p. 192). Analogously, Westermarck reasoned, should there be a 'natural' incestuous inclination, a prohibition against acting out such an inclination it should not extinguish the inclination itself.

However, if there is no inclination to engage in incest, why is it condemned in custom and law? Questioning Westermarck's explanation, the Marxist French anthropologist Maurice Godelier (1989) delivered a possible explanation which, somewhat ironically, was in line with contemporary sociobiological thought. Concerning rules against incest, he suggested that when the social conditions have supported the practice of incest, only religious prohibitions were able to control it. This could suggest that, *whenever* there is an inclination towards incest, social norms divert this inclination and obstruct its manifestation. The position of Frazer, Freud and their followers, and the opposing position of Westermarck, both seem rather categorical – the former suggesting incestuous inclinations to be common, and the latter suggesting incestuous inclinations to be inexistent. Perhaps there is variation in its presence. Pertaining to this, David Haig (1999) argued that there is a biological reason for an asymmetry between family members regarding the motivation to avoid inbreeding. His model builds on inclusive-fitness theory, showing that the cost of incest to an individual depends on a number of factors. For example, according to Trivers' (1972) theory describing gender differences in parental investment, women tend to have a higher cost of inbreeding than do men. In the case of conception, women are pregnant for nine months followed by a lengthy period of lactation, during which oestrus is generally suppressed. This means that women lose many opportunities to engage in more optimal breeding. Men, on the other hand, lose on average fewer opportunities, as their minimum investment is far less. This leads to the expectation that women should be more averse to incest than men, something that a number of studies has shown over a variety of measures (e.g., Antfolk et al., 2012; Antfolk, Lieberman, & Santtila, 2012; Fessler & Navarrete, 2004; Lieberman, Tooby, & Cosmides, 2003). Furthermore, as Westermarck was the first to point out, the mechanism activating human inbreeding avoidance is imperfect, leading to a number of false positives and false negatives. This suggests that there are instances where one could expect that, for example, a father does not feel aversion towards a daughter, although the daughter feels aversion towards him. In such instances, a social proscription against acting out such inclinations – a cultural incest taboo – could decrease the likelihood of an incestuous act occurring. Studies have suggested that such prescriptions could be evolutionarily crafted, although the prevalence of incestuous inclinations would be rather low (Aoki, 2004). However, although this possibility is necessary for an evolutionary explanation, it can by no means be considered as proof thereof.

Although the assumption of the social extension of the individual aversion has received a considerable amount of academic interest, very few empirical studies have been aimed at testing this assumption. Constructing a direct experiment that tests this assumption seems hopeless. However, in a few studies researchers have asked individuals who have lived together with an opposite-sex sibling in childhood – a factor expected to increase individual aversion – to describe their reaction towards written descriptions of *others* engaging in incest. These studies have generally found that in studies where participants have been asked about the moral wrongness of a number of acts, individuals coresiding with an opposite-sex sibling in childhood have ranked incest as more morally wrong than other individuals (Lieberman et al., 2003). Also, individuals who have coresided with an opposite-sex sibling find incest more aversive (Fessler & Navarrete, 2004).

However, there are some methodological limitations to these studies. First, they do not show that the increased aversion to third-person inbreeding descriptions is mediated by an increased aversion to the participants themselves engaging in incest. Second, this type of pen-and-paper studies has questionable ecological validity. For example, the interpretations of various statements and descriptions employed in these types of studies may vary from the interpretations of real-life situations. Further, the degree to which a fictive incest story is aversive in comparison to a *real* incest case is unknown. Moreover, they cannot rule out that individuals who have coresided with an opposite-sex sibling are more likely to intentionally comply with an existing incest taboo. Incest simply may not be such an important matter to those who have no siblings. However, Westermarck's assumption predicts that if the social taboo is an extension of personally felt aversions, the degree to which we feel indignation towards *others* engaging in incest should covary on the costs we should suffer in a similar situation in which *we* participated. The Frazer–Freud assumption that implies an already existing taboo that each individual internalises does not parsimoniously explain an association between fitness costs to one's self and the strength of indignation felt towards others engaging in incestuous activities. In line with Westermarck's reasoning, a recent study by Antfolk et al. (submitted) showed that individual aversion to incest scenarios was closely predicted by fitness costs, and this pattern is present irrespective of whether the participant himself or herself is described as participating in the incest or not. More exactly, participants' responses regarding individuals related or unrelated to the participant engaging in incest were predicted by the fitness decrease in the offspring from such unions. However, as pointed out by the authors, the fitness costs implied to the participants are not the only possible explanation. It is possible that because there is a cultural expectation that an individual *should* feel more responsible to intervene when close kin are at risk. Thus, it is plausible that some other factor, such as a stronger emotional bond, covaries with both aversion and degree of relatedness and thus introduces a spurious association. Pertaining to this, Ackerman, Kenrick and Schaller (2007) compared reactions to sexual activities with a close friend with reactions to sexual activities with a sibling. This study showed that, rather than emotional closeness, relatedness was associated with the magnitude of aversion.

Much like the three other assumptions embedded in Westermarck's explanation of human incest avoidance, the assumption that the incest taboo is a social extension of individually felt aversions has not received compelling support. Neither has it been falsified. No study has yet been designed that could comprehensively separate between the explanations provided. Further, there is something of an internal contradiction between incest aversion being a social extension and the other assumptions. Should Westermarck be correct in that coresidence during childhood leads to an inbreeding aversion and that *this* aversion extends into social norms, we should not expect to see societies in which the one exists without the other. In fact, we have seen that there *are* situations in which the culture allows future partners to coreside during their early childhood, even societies that *expect* future partners to coreside during their childhood. Studies, however, also show an increased moral indignation towards sexual unions between individuals that have coresided, even when these individuals are not biologically related (Antfolk et al., 2012).

Conclusions

In the late 19th century, Westermarck presented his hypothesis regarding human incest aversion. His hypothesis was initially warmly received, only to fall into disfavour just a few decades later. If Westermarck's hypothesis was accepted and rejected in an empirical vacuum, its concordance with the current evolutionarily understanding of the human family system can be illuminated by some, however incomplete, empirical data. Although there is some support for the expected effect of coresidence in childhood on incest aversion, it seems probable that coresidence alone cannot be the only agent moderating incest aversion. Rather, coresidence is thought to be one of several available cues that regulate the certainty in biological relatedness, which is thereby also associated with incest aversion. Thus, in understanding human incest aversion, Westermarck's hypothesis may be a part of the solution, but may not provide the full solution.

Although Westermarck's explanation more parsimoniously explains the available data than the psychoanalytical or cultural explanations, direct, thorough and convincing support is still lacking. Because explaining the apparent lack of sexual desire between close kin and explaining the origin of the culturally universal incest taboo is at the crossroad of a number of historically important discourses pertinent to human behaviour, it is likely that the debate will continue and that scholars will attempt to subject their positions to more and more rigorous tests. Until such rigorous tests have been conducted, we cannot know for certain to which degree Westermarck was right.

8 The Westermarck thesis as a thinking tool for sociobiology

Ullica Segerstrale

With his discovery of the seemingly universal reluctance of mating between individuals who had been raised together (the so-called Westermarck effect), Edward Westermarck was the first sociologist to bring a type of Darwinian reasoning into sociology and anthropology that foreshadowed the thinking in current sociobiology and evolutionary psychology. His thesis, although first very well received, later came into conflict with other explanations, especially the dominant Freudian and functionalist anthropological views (e.g., Allardt, 2000). After a long time of invisibility, Westermarck's thesis has been revived, beginning in the last quarter or so of the 20th century. Especially for Edward O. Wilson, who in 1975 launched the field of sociobiology, Westermarck's thesis has taken on a dual role of a 'thinking tool' and a vivid example in his expanding sociobiological reasoning over the years. This chapter follows Wilson's engagement with Westermarck's thesis – and especially the Westermarck effect – in Wilson's books and articles from 1975 to 2012 as he has been taking on new research themes. In particular, this chapter examines the particular role the Westermarck's effect has been given on different points in his evolving sociobiological programme.

Westermarck's thesis consists of several claims, and people do not necessarily agree about how they relate to one another. In the first place, how many 'parts' of the hypothesis are there? The philosopher Neven Sesardic (1998, 2005) identified three different claims:

1. A sexual aversion tends to develop between those raised together in early childhood.
2. This aversion is an evolutionary adaptation, serving as a barrier to inbreeding depression (a drastically decreased biological viability of offspring from unions between close biological kin).
3. The aversion causes (expresses itself as) the incest prohibition (a cultural prohibition against incest or an incest taboo).

After Durham (2005), these claims can be called the aversion hypothesis, the adaptation hypothesis and the expression hypothesis. The first one says that the phenomenon of sexual aversion exists, the second one that its cause is natural selection and the third one that its effect is the incest prohibition. (For an alternative view of the 'parts' see Antfolk and Wolf, Chapter 7 in this volume.)

Ever since Westermarck published his original book in 1891 these three claims have been supported, separately or together, and also challenged. This has typically been done by social scientists (anthropologists, sociologists and Freudians), a trend that continues today. What critics identify as major flaws, however, can often be routinely accommodated within an evolutionary framework (for instance, the point that a behaviour does not have to be expressed 100% to be evolutionarily grounded).

Durham (2005) and Sesardic (2005) agree that the parts of the Westermarck hypothesis that have the strongest support are the first two, whereas the 'expression hypothesis' is the weakest link.

So how do the different parts hold together in Westermarck's own formulation? Here is a comprehensive quote of what Westermarck actually said (as he was commenting on contemporary views of anthropologists and Freudians):

> [T]he theories in question imply that the home is kept free from incestuous intercourse by law, custom, or education. But even if social prohibitions might prevent unions between the nearest relatives, they could not prevent the desire for such unions. The sexual instinct can hardly be changed by prescriptions [. . .]. Nevertheless, our laws against incest are scarcely felt as a restraint upon individual feelings. And the simple reason for this is that in normal cases there is no desire for the acts they forbid. *Generally speaking, there is a remarkable absence of erotic feeling between persons living closely together from childhood*; among the lower animals, also, there are indications that the pairing instinct fails to be stimulated by companions and seeks strangers for its gratification. [. . .] Plato showed a sharper eye for the problem of incest in his observations that an unwritten law defends as sufficiently as possible parents from incestuous intercourse with their children and brothers from intercourse with their sisters, and that the thought of such a thing does not enter at all into the minds of most of them.
>
> *Sexual indifference, however, is not by itself sufficient to account for exogamous prohibitions. But such indifference is very generally combined with sexual aversion when the act is thought of*; indeed, I believe that this is normally the case whenever the idea of sexual intercourse occupies the mind with sufficient intensity and a desire fails to appear. [. . .] [*A*]*versions which are generally felt readily lead to moral disapproval and prohibitory customs and laws*. This I take to be the fundamental cause of the exogamous prohibitions. Persons who have been living together since childhood are as a rule near relatives. Hence their aversion to sexual relations with one another displays itself in custom and law as a prohibition of intercourse between near kin.
>
> (Westermarck, ER, pp. 247–9, cited in Arnhart, 2005a, pp. 200–1; italics added)

For a convinced Darwinian thinker, there is no particular mystery involved. Westermarck himself saw natural selection as the motor behind everything. Originally he briefly wrote about an 'instinct' developed to avoid incest:

> Through natural selection an instinct must have been developed, powerful enough, as a rule, to prevent injurious unions. This instinct displays itself simply as an aversion on the part of individuals to union with others with whom they have lived, but as these are for the most part blood-relations, the result is the survival of the fittest.
>
> (Westermarck, HHM, p. 546)

This was Westermarck himself. We will now follow Wilson's engagement with the Westermarck thesis and the issues it addresses.

Sociobiology as the new synthesis: inbreeding avoidance as a tool for survival

In *Sociobiology*, Wilson addresses the issue of incest largely from the perspectives of genetics, population genetics and ethology. He starts by discussing the phenomenon of inbreeding depression and the observed detrimental biological outcome of incestuous unions in humans. Also, with inbreeding the diminished genetic variability of the offspring would make them less adaptable to change. Inbreeding, Wilson (1975) argues, would therefore 'threaten the very survival of a group'. Wilson notes, however, that were it not for the deleterious results of inbreeding, inbreeding would, in principle, be a way for individuals to increase their inclusive fitness. 'Presumably, then, the degree of sociality is to some extent the evolutionary outcome of these two opposed selection tendencies', Wilson muses, and suggests that modelling this will be an important challenge for population genetics (Wilson, 1975, p. 80).

Animals have a number of behavioural mechanisms to avoid inbreeding, such as young animals leaving or being chased away, but we have to go human beings to find behaviour patterns uniquely associated with incest taboos. The most basic process appears to be the process that Tiger and Fox (1971) have called 'the precluding of bonds'. Wilson notes and explains that 'fathers and daughters, mothers and sons, and brothers and sisters find their primary bonds to be all-exclusive, and incest taboos are virtually universal in all human cultures' (Wilson, 1975, p. 79). He then goes on to show that the bond exclusion is not dependent on biological relatedness, but has rather to do with close domestic proximity. He brings up Joseph Shepher's (1971) Israeli *kibbutz* study, which showed 'that [sexual] bond exclusion among age peers is not dependent on sibship' (Wilson, 1975, p. 79). Shepher found that of 2,769 marriages not a single one was between individuals from the same *kibbutz* who had been living together since birth. Meanwhile, there were no formal or informal pressures to refrain from marrying peers.

In *Sociobiology*, an encyclopaedic work, the ideas of incest and inbreeding avoidance are not allowed much space. Wilson accepts his biosocial anthropologist colleagues' view of the incest taboo as a result of strong primary bond formation, and in relation to sociobiology he sees the incest issue as primarily involving a population genetic modelling challenge. Westermarck and the Westermarck effect are not mentioned.

On human nature: inbreeding avoidance as a human universal and example of 'the genes hold culture on a leash' argument

Wilson's following book, *On Human Nature* (1978), a popularly written book, won the author a Pulitzer Prize. It was an expansion of the controversial last chapter in *Sociobiology* and included discussions of values, morals and religion as having a biological underpinning. But his primary aim in this book was to establish the existence of a universal human nature – whose very existence was doubted by his ferocious critics (for the sociobiology debate, see Segerstrale, 2000).

A good way of bringing home the idea of human universals is to involve the existence of incest taboos. Wilson is at the same time appealing to the general knowledge and the intuition of his readers:

> Incest taboos are among the universals of human social behavior. The avoidance of sexual intercourse between brothers and sisters and between parents and their offspring is everywhere achieved by cultural sanctions.
>
> (Wilson, 1978, p. 36)

But then he continues, pointing to the importance of proximity:

> But at least in the case of the brother-sister taboo, there exists a far deeper, less rational form of enforcement: a sexual aversion automatically develops between persons who have lived together when one or all grew up to the age of six.
>
> (Wilson, 1978, p. 36)

He again notes that studies in Israeli kibbutzim, 'have shown that the aversion between people of the same age is not dependent on an actual blood relationship', but now goes on to explain how the system is actually working. Both proximate and ultimate causes are involved:

> The automatic exclusion of sexual bonding between individuals who have previously formed certain the kinds of relationships – the 'gut feeling' that promotes the ritual sanction against incest – is largely unconscious and irrational. Bond exclusion of the kind displayed by Israeli children is an example of what biologists call a proximate (near) cause; in this instance, the direct psychological exclusion is the proximate cause for the incest taboo. The ultimate

> cause suggested by the biological hypothesis is the loss of genetic fitness that results from incest. [. . .] Natural selection has probably ground away along these lines for thousands of generations, *and for that reason human beings intuitively avoid incest through the simple, automatic rule of bond exclusion.* To put the idea in its starkest form, one that acknowledges but temporarily bypasses the intervening developmental process, human beings are guided by an instinct based on genes.
>
> (Wilson, 1978, p. 38, italics added)[1]

Wilson again refers to *kibbutz* studies and Shepher's (1971) research on mate selection among *kibbutz* peer groups, but the Westermarck thesis is not mentioned in the text. But when we go to the endnotes for the chapter, after the citation to Shepher's study, we find an added note: 'The possibility of automatic aversion based on early domestic intimacy was first suggested by Edward Westermarck in 1891' (Wilson, 1978, p. 229, note 36).

Wilson goes on to explain the existence of incest taboos. He dismisses the typical anthropological explanations of why they exist, such as the need for preserving family roles and the integrity of the family, or the facilitation of exchange of women during political bargaining with other groups, suggested by Edward Tylor. He contrasts these types of theories with 'the prevailing sociobiological explanation', which regards those reasons as merely contributing factors or by-products, whereas the prevailing sociobiological explanation 'identifies a deeper, more urgent cause, the heavy physiological penalty imposed by inbreeding'. Inbreeding vastly increases the risk of existing deleterious recessive genes to combine in a homozygous state, causing death, severe mental retardation and various deformities and abnormalities in the offspring, he warns. Moreover, geneticists have shown that even a moderate amount of inbreeding already causes pathologies in the offspring (Wilson, 1978, p. 37).[2]

He concludes with the following provocative statement:

> The manifestations of inbreeding pathology constitute natural selection in an intense and unambiguous form. The elementary theory of population genetics predicts that *any behavioral tendency to avoid incest, however slight or devious, would long ago have spread through human populations. So powerful is the advantage of outbreeding that it can be expected to have carried cultural evolution along with it.*
>
> (Wilson, 1978, pp. 37–8, italics added)

This is why those various advantages that anthropologists have attributed to outbreeding are more likely to be 'secondary cultural adaptations that make use of the inevitability of outbreeding for direct biological reasons', Wilson observes (Wilson, 1978, p. 38). But he is here not only dismissing functionalist anthropological explanations, he is also foreshadowing his own later formulation in *On Human Nature*: '[t]he genes hold culture on a leash' (Wilson, 1978, p. 167).

Genes, mind and culture: the Westermarck effect as an epigenetic rule

The obvious next step for Wilson would be somehow scientifically to demonstrate that genes could, in fact, hold culture on a leash. And this was, indeed, his ambition in his following book *Genes, Mind, and Culture*, co-written with Charles Lumsden, who was trained in theoretical physics but eager to do modelling in biology. A mathematical modeller was exactly what Wilson needed at this point.

The project would be to put the reasoning about humans in an explicit mathematical form. The critics of sociobiology had persistently claimed that population genetics was not able to explain cultural differences. It required too long a time span. They doubted Wilson's claim in the last chapter of *Sociobiology* that the small genetic differences that admittedly existed between human populations could have given rise to the great cultural diversity of humans in just a few thousand years. But population genetics was exactly the tool that Wilson wanted to use in order to bring social behavior into the modern synthesis. The specific point therefore was somehow to demonstrate mathematically how Wilson's initial insight could in fact work for humans.[3]

The result of the collaboration was a complete change of thinking in regard to sociobiology, a move away from the earlier sociobiological idea that genes directly prescribed behaviour to a new vision of gene–culture coevolution. Rather surprisingly, in the new book Lumsden and Wilson themselves declared the original kind of sociobiological reasoning incorrect. This was the way the duo upfront formulated their new position, having moved from sociobiology I to sociobiology II, as it were:

> Behavior is not explicit in the genes, and mind cannot be treated as a mere replica of behavioral trait. In this book we propose a very different view in which the genes prescribe a set of biological processes, which we call *epigenetic rules*, that direct the assembly of the mind. This assembly is context dependent, with the epigenetic rules feeding on information derived from culture and physical environment [. . .] *[C]ulture is the translation of the epigenetic rules into mass patterns of mental activity and behavior* [. . .] Genes are indeed linked to culture, but in a deep and subtle manner.
>
> (Lumsden & Wilson, 1981, p. 2; italics added)

According to the authors, the 'epigenetic rules' affect the choices of individuals when it comes to 'culturgens' (cultural behaviours, concepts, ideas or artefacts). Culturgens are seen as isolated units of culture that individuals may choose to adopt or not adopt. But this choice is not totally free: it is constrained by pre-existing epigenetic rules, or biases of the mind. 'The epigenetic rules will [. . .] tend to channel cognitive development toward certain culturgens as opposed to others. We refer to this relation informally as the "leash principle" in order to make it metaphorically more vivid: *genetic natural selection operates in such a way as to keep culture on a leash*' (Lumsden & Wilson, 1981, p. 13, italics added).[4]

The epigenetic rules guide the process by which the genes are expressed during development and constrain the operation of the mind. The primary ones are more inflexible, biasing sensory input (e.g., colour perception, hearing, taste, smell); the secondary ones bias evaluation and decision making (Lumsden & Wilson, 1981, p. 52). Behaviours guided by the primary epigenetic rules are incorporated automatically in most or all human societies – they are cultural universals. So what would be a good example to show how this new genes-mind-and-culture model works? Besides colour perception, the authors' favourite example in *Genes, Mind, and Culture* is incest avoidance between siblings, which they also use to demonstrate their mathematical modelling. Their incest avoidance example was also the subject of a paper in the Proceedings of the National Academy in 1980 (Lumsden & Wilson, 1980).

Where is Westermarck mentioned in this book? Let's see. In the discussion of brother–sister mating in *Genes, Mind, and Culture*, the authors mention the various explanations for the incest avoidance given by Freud and the anthropologists, as well as a recent cross-cultural study by Ember (1975), which suggests that the observed consequences of inbreeding could indeed have given rise to deliberately created incest taboos. But they immediately continue:

> However, analyses by Shepher (1971) and others on the development of sexual preferences of children in Israeli kibbutzim indicate that this is not the case. *There exists instead a relatively specific epigenetic rule*, whereby an automatic sexual inhibition between persons who lived intimately together ("used the same potty") emerged as one or all grew to the age of six.
>
> (Lumsden & Wilson, 1981, p. 86, italics added)

Lumsden and Wilson go on to mention the absence of marriages in this case and then bring in the 'closely parallel data' provided by Wolf and Huang of Taiwanese minor marriages. Whereupon Lumsden and Wilson conclude: 'The evidence suggests the existence of a genetically based bias curve in which the preference for outbreeding as opposed to incest is very strong' (1981, p. 86). We are then informed that this behaviour will be more closely analysed in Chapter 4 and used as an example of the translation from epigenetic rules to cultural patterns.

But in Chapter 4, under the subheading Brother-Sister Avoidance, we are just met with a summarising statement: '*The epigenetic rule appears well established*: a deep sexual inhibition develops between people who live in close domestic contact during the first six years of life'. Later in that chapter we learn:

> The avoidance of sibling incest is generated by a strong epigenetic rule in which close domestic familiarity during the first six years of life neutralizes sexual attraction [. . .]. This rule has been verified by studies in Israel and Taiwan.
>
> (Lumsden & Wilson, 1981, p. 148, italics added)

After this follow details about Shepher's and Wolf's empirical studies.

But whose *is* this rule? 'There exists instead a relatively specific epigenetic rule' – but where does it come from? (Chapter 3 turned out not to be helpful). Joseph Shepher is cited, and so is Arthur Wolf, and their research is being described as verifying the rule. But there is no reference to Westermarck. This is a surprise. Perhaps the authors consider theirs a purely empirically based book, which is why they cite only empirical findings, rather than other people's theories? Well, but they do cite the theories for example of Freud and the anthropologists. Moreover, it is clear that at least Wolf saw himself as explicitly verifying Westermarck's thesis. His 1970 paper, cited by the authors, had the intriguing title, *Childhood Association and Sexual Attraction: A Further Test of the Westermarck Hypothesis*. And we know that Wilson knows about Westermarck. This is puzzling. We will just continue and see if an explanation suggests itself.

Promethean fire: the Westermarck thesis as a tool for social engineering

The situation gets even stranger with the next book by Lumsden and Wilson. *Promethean Fire* (1983) is a popular book with the aim to explicate many of the technical sounding points in *Genes, Mind, and Culture*. It is rich with examples – and guess which one is the most salient? In fact, the book's two final chapters are devoted to an intimate discussion of brother–sister incest avoidance. It involves even the imagined reasoning of a tribesman considering the possibility of incest with his sister (reportedly reconstructed from anthropological interviews).

At the time, Lumsden and Wilson were very proud of their newly developed model of gene–culture coevolution, in which genetic evolution was coupled to cultural evolution in a mutual feedback process (Segerstrale, 2000). Note that with its internal coupling, the model was rather different from the coevolutionary model earlier developed by the anthropologist Bill Durham (1978). Durham's model assumed relative independence between genetic and cultural evolution, recognising that cultural beliefs and practices could independently have important fitness consequences for a particular human population. But in *Promethean Fire*, Lumsden and Wilson pointed out an unexpected great advantage of their own more tightly 'coupled' theory. Knowledge about the link between genes and culture could be valuable for social planning or, rather, social engineering!

The authors painted a bold vision: 'Social engineering has the potential of profoundly altering every part of human behavior [. . .] Some very human propensities, which may have been of great adaptive value in the Stone Age, are now largely self-destructive'. Societies would be able to employ knowledge of the epigenetic rules 'to guide individual behavior and cultural evolution to the ends on which their members may someday agree'. For instance, some undesirable propensities might be blunted (aggression, xenophobia) and desirable ones enhanced. At the social level again, '[t]he value of institutions and forms of government can be more accurately judged, alternative procedures lain out and steps carefully suggested' (Lumsden & Wilson, 1983, pp. 183–4).

What, then, would serve as a good example of this kind of hypothetical social engineering, based on our knowledge of the epigenetic rules? The authors have one handy example already: brother–sister incest! Let's say a society would like to promote brother–sister incest, they propose. How would it go about promoting it? The answer would be to use the knowledge we have about the epigenetic rules to manipulate the environment. We know that children raised together during the first six years or so are not physically attracted to each other. So, they suggest, society might simply arrange for sisters and brothers to be raised apart. At the same time, the culture should in different ways celebrate the idea of sibling romance and marriage (Lumsden & Wilson, 1983, pp. 176–7).[5]

Why would this work, according to the authors? Lumsden and Wilson draw a parallel between their social engineering idea and a well-known example: the way in which it is possible to correct the genetic defects of phenylketonuria by the right diet. *Through close knowledge of gene–environment interaction it is possible to select the precise environment that reverses the usual response* – be it phenylketonuria or sister–brother incest, they note (Lumsden & Wilson, 1983, p. 177). The authors have an even larger vision of the new possibilities now opening up:

> *A sufficient knowledge of genes and mental development can lead to the development of a form of social engineering that changes not only the likelihood of the outcome, but the deepest feelings about right and wrong, in other words, the ethical precepts themselves.*
>
> (Lumsden & Wilson, 1983, p. 179, italics added)

This is strong stuff.

This foreshadows the next instalment in our ongoing story, in which the Westermarck effect becomes the prize example of a moral instinct. But let's still stick with *Promethean Fire*. What happened to naming Westermarck in that book? The theme of sibling incest is introduced on page 64, backed up by the usual description of the *kibbutz* and minor marriage studies. This time, however, no names at all are mentioned – not Westermarck, not Shepher, not Wolf. We see no references to these individuals and their studies either in an accompanying footnote, which states: 'The brother-sister incest case is discussed further, with documentation, in the final two chapters of this book' (Lumsden & Wilson, 1983, p. 195). So we go there to find more information.

The penultimate chapter contains, among others, an imagined scenario where a tribal member, Champukwi, thinks about incest with his sister and imagines what it would be like. 'He winces', we learn (Lumsden & Wilson, 1983, p. 126). The presentation continues with a sketch of the fluctuating sexual preferences of the larger tribal community. On any given day, it is assumed by the model, these depend on the choice between two culturgens in the minds of each ruminating individual, producing a statistical result for the tribe as a whole (Lumsden & Wilson, 1983, p. 127). Nevertheless (presumably because of the influence of an epigenetic rule biasing the decision) the choice for incest avoidance is estimated to be high, around 90%.

We meet the brother–sister incest case again on pages 133–4 and are given details about the usual empirical studies, whose authors this time remain nameless. However, we are told, '*[t]he dominant epigenetic rule appears to be strong and clear-cut*: when children are raised together in close domesticity during the first six years of life they are automatically inhibited from full sexual activity at maturity' (Lumsden & Wilson, 1983, p. 133, italics added). Moreover, we learn that the preference for brother–sister incest was deliberately chosen as an extreme example to help Lumsden and Wilson in their modelling. It permitted them 'to devise the first estimates for a real case of translation from individual mental development to cultural pattern' (Lumsden & Wilson, 1983, p. 134).[6]

The Westermarck effect as "a more robust technique of moral reasoning"

There are, however, yet bigger ambitions for the epigenetic rules. The next instalment in our story is Wilson's 1986 paper written together with the philosopher Michael Ruse and entitled 'Moral Philosophy as an Applied Science'. This collaboration directly follows up one of Wilson's provocative suggestions in *Sociobiology*: that ethics ought to be taken away from the philosophers and biologised. Ruse and Wilson aim to show that morality does not belong to a separate realm of culture but is empirically grounded:

> The empirical heart of our discussion is that *we think morally because we are subject to appropriate epigenetic rules*. These predispose us to think that certain courses of action are right and certain courses of action are wrong [. . .] under the influence of the rules certain choices are made from among those conceivable and available to the culture; and finally the choices are narrowed and hardened through contractual agreements and sanctification.
>
> (Ruse & Wilson, 1986, pp. 180–1, italics added)

The authors want to show that just like various types of cognition, moral reasoning, too, is moulded and constrained by epigenetic rules. For this they need a convincing example. Yes, there is one: the avoidance of brother–sister incest (Ruse & Wilson, 1986, p. 183)! After a brief overview of the negative effects of incest and their genetic explanation, they note that the cause of the avoidance is a sensitive period between birth and approximately six years for children raised in close proximity. Here they bring in the *kibbutz* and Chinese minor marriage studies, citing Wolf and Huang (1980) and Shepher's then recent 1983 book. They now continue:

> *A widely accepted interpretation of the chain of causation in the case of brother-sister incest avoidance is as follows*. Lowered genetic fitness due to inbreeding led to the evolution of the juvenile sensitive period by means of natural selection; the inhibition experienced at sexual maturity led to prohibitions and cautionary myths against incest or (in many societies) merely a

> shared feeling that the practice is inappropriate. Formal incest taboos are the cultural reinforcement of the automatic inhibition, an example of the way culture is shaped by biology. But these various surface manifestations need not be consulted in order to formulate *a more robust technique of moral reasoning. What matters in this case is the juvenile inhibition*; the measures of its strength and universality, and a deeper understanding of why it came into being during the genetic evolution of the brain.
>
> (Ruse & Wilson, 1986, p. 184, italics added)

This time the case of brother–sister incest avoidance is presented as a 'widely accepted interpretation of the chain of causation' linking culture to biology, according to which formal incest taboos represent 'the cultural reinforcement of the automatic inhibition', as well as the already mentioned 'more robust technique of moral reasoning'. And again the authors find an even grander use for this example. It can be upheld to refute the very view that morality is a product of culture:

> The incest example shows that [the epigenetic] rules, directly related to adaptive advantage, extend into the moral sphere. *And the hypothesis of morality as a product of pure culture is refuted by the growing evidence of the co-evolution of genes and culture.*
>
> (Ruse & Wilson, 1986, p. 185, italics added)

The authors venture yet further. *There can be no genuinely objective external ethical premises!* This is because the ethical premises of the human species arise from the particular epigenetic rules of mental development, and those are idiosyncratic products of the species history, shaped by natural selection. *Our ethical principles are uniquely linked to the fact that we are human* (if we were, say, termites, we would favour completely different principles). 'No abstract moral principles exist outside the particular nature of individual species [. . .] Morality is rooted in a contingent human nature, through and through' (Ruse & Wilson, 1986, p. 186, italics added). This is about as far as one can push the implications of the Westermarck thesis.

Now, where does Westermarck enter in all this discussion? He doesn't. He is the invisible man.

Consilience: the Westermarck thesis as a bridge between the two cultures

The most ambitious project from a philosophical point was surely Wilson's book *Consilience* published in 1998.[7] 'Consilience' was Wilson's term for the possibility and need for the unification of the 'two cultures', a long-standing theme of his. The challenge was to find explanations of the kind that would be able to span the gap between nature and culture, the natural sciences and the liberal arts. On the face of it, it was a tall order, but Wilson believed that he had found the answer. The right tools for mediating between nature and culture were to be found in the general principle of gene–culture coevolution and the workings of the epigenetic rules. The epigenetic rules would help him make sense of the world as a 'seamless web of cause and effect', the ultimate goal of his project (Wilson, 1998, p. 188).

For this argument to work, an important point would be to prove that widely distributed cultural traits – cultural universals – are typically adaptive (another way of saying that the genes hold culture on a leash). Wilson admitted that there was not a lot of concrete evidence for this 'genetic fitness hypothesis'. But there was *one* good example that supports the hypothesis: incest avoidance!

> *The category of human behavior that provides the fullest test of the genetic fitness hypothesis to date is incest avoidance.* A large account of information concerning the phenomenon has become available at different levels of biology and culture. The behavior itself is universal, or nearly so. It is also relatively clear-cut in expression. Sexual activity in all societies is relatively uncommon between siblings and between parents and their offspring; children produced by such activity are rare; and long-term unions made with the consensual purpose of having such children are almost nonexistent.
>
> The current explanation of incest avoidance, which combines genetic and cultural evolution, is a relatively straightforward sociobiological exercise. Inbreeding at the level of siblings and parents yields a high percentage of offspring with genetic defect. Humans tend to avoid this risk by unconscious obedience to the following epigenetic rule: *If a boy and a girl are brought together before one or the other is thirty months of age*[8] and then raised in close domestic proximity [. . .] they are devoid of later sexual interest in each other, and the very thought of it arouses an acute aversion. This emotional incapacity, fortified in many societies by a rational understanding of the consequences of inbreeding, has led to the cultural incest taboos, which prohibit incest by custom and law.
>
> (Wilson, 1998, p. 188, italics added)

Wilson goes on to describe the risk of inbreeding depression and the ways in which this is avoided in many animals, especially primates, and in plants, too. Primates leave the home group and join another before they reach sexual maturity; this is reminiscent of human exogamy (the principle of marrying outside one's group). For those who remain in the group, again, there is a 'second line of resistance' against inbreeding: to avoid mating with each other. Wilson notes that in all the nonhuman primate species that have been studied, both males and females spurn individuals with whom they grew up (Wilson, 1998, pp. 189–90). This time, however, at the end of his biological lecture Wilson brings in Edward Westermarck and gives him full recognition:

> This elemental response was discovered, not in monkeys and apes, but in human beings, by the Finnish anthropologist Edward A. Westermarck and first reported in his 1891 masterwork *The History of Human Marriage*. The

> existence of the phenomenon has gained increasing support from many sources in the intervening years.
>
> (Wilson, 1998, p. 190)

This time Wilson's stance in regard to the existing evidence seems bolder. He says he considers the case of the *Sim-Pua*, Taiwanese minor marriages, studied by Arthur Wolf during more than 40 years, 1957–1995, to be the most persuasive. 'The results unequivocally favor the Westermarck hypothesis', Wilson confidently notes, as he comments on Wolf's meticulously controlled research involving various cross-analyses to identify and eliminate other possible factors that could have played a role in the aversion to mating between coraised children. After citing Shepher's *kibbutz* studies and other anecdotal evidence, he declares: 'It is evident that the human brain is programmed to follow a simple rule of thumb: *Have no sexual interest in those whom you knew intimately during the earliest years of your life*' (Wilson, 1998, p. 191, italics in original).

More support for the reality of the Westermarck effect, according to Wilson, comes from the fact that it is 'graded' – that is, the more intimate the early childhood association, the stronger is the later aversion to mating. Although 'we are still far from a full explanation of incest avoidance', Wilson thinks that 'everything indicates that the incest avoidance effect is genetic'. He can even explain why the effect is so universal. Only very small genetic differences would have been needed for it to spread:

> Given even a small amount of genetic variability in sexual responsiveness to childhood associates, the differences in fitness based on it would have been strong enough, in population genetic theory at least, to spread the Westermarck effect throughout the population from a very low incidence to widespread occurrence in as few as ten generations.
>
> (Wilson, 1998, pp. 191–2)

Wilson concedes, though, that the psychological mechanism of the Westermarck effect – the critical stimulus that triggers the aversion reaction – remains unknown. We know, however, that simple cues can release or inhibit quite complex behaviours; this is the way instincts work. It could be something very small and unconscious associated with two individuals' close past association, such as a touch or a scent at a critical moment, Wilson muses.

There is still a 'third line of resistance': the incest taboos. These taboos are conscious inventions, Wilson explains, often interwoven with principles of kinship classification and exogamous marriage contracts. In contrast to the taboo on sibling marriage, many societies in fact encourage marriage between first cousins, he adds, because this often promotes group cohesion or wealth consolidation (Wilson, 1998, p. 192).[9]

But what about the known exceptions to the incest rules? This is the first time that Wilson addresses this matter in detail. The exceptions include the Incas, Hawaiians, Thais, ancient Egyptians, Nkole and Bunyoro and Ganda (Uganda),

Zande (Sudan) and Dahomeyans (West Africa). In these incestuous arrangements – typically in ruling families – the males mated also with genetically unrelated females, Wilson observes. So in this way they could guarantee not only high inclusive fitness – being 75% related to their offspring – but also healthy offspring. A well-documented exception – brother–sister marriage among commoners, rather than ruling families – was that of Roman Egypt, where it was practiced for a considerable time: 30 B.C. to A.D. 324 (Wilson, 1998, p. 193).

The Westermarck thesis as a weapon against Freud

Wilson's grand aim in *Consilience* was to bring together the two cultures and connect them in a causal way. We saw how Wilson employed the Westermarck effect to prove that this was possible in principle. Indeed, as he observed himself: '[i]ncest taboos have led us, once again, to the borderland between the natural and social sciences' (Wilson, 1998, p. 193). But there is still the need to defend the Westermarck thesis against the main intellectual contender: the Freudian thesis of the cultural origin of the incest taboo. Wilson prepares himself for battle. *What is then the relation between the Westermarck effect, which is biological, and the incest taboos, which are cultural*, he provocatively asks? (Wilson, 1998, p. 193).

Wilson does a quick recap. Westermarck said (in Wilson's 'updated language', as he calls it): '*people avoid incest because of a hereditary epigenetic rule of human nature that they have translated into taboos*' (italics added). For Freud, on the other hand, the primal lust for sex between family members was so strong that society needed to construct taboos to sustain the integrity of the family. According to him, the cultural taboo was invented to suppress the natural biological instinct of incest. Not only did Freud declare Westermarck's idea preposterous, but he argued, just like the anthropologist James Frazer before him, that if the Westermarck principle was correct, there would be no need for a taboo. Freud 'won', and his view became the orthodoxy for most of the rest of the century. But Westermarck had actually responded already to Frazer, before Freud. Here is Westermarck's argument, in popular reconstruction by Wilson:

> Individual humans, Westermarck said, reason as follows. *I am sexually indifferent to my parents and siblings. Yet I occasionally wonder what it would be like to have sex with them. The thought is repugnant! Incest is forced and unnatural. It would alter or break other bonds I have formed with them and must maintain on a day-to-day basis for my own welfare. Incest by others is by extension also repugnant to my mind, and evidently to that of others too, and so the rare cases in which it occurs should be condemned as immoral.*
>
> (Wilson, 1998, p. 194, italics in the original)

According to Wilson, Westermarck's response to Frazer was 'simple, equally logical, and supported by growing amounts of evidence'. Wilson also agrees with Wolf (1995) that Freud saw clearly that if Westermarck was right, *he* was wrong, which would make all his psychoanalytic theory, with the Oedipus complex and the origin of neuroses – not to speak about his grand explanation of the origin of civilization itself – come tumbling down. Westermarck, meanwhile, had been arguing in his more recent writings on ethics that moral concepts are derived from human emotions, Wilson reports. This undermined the traditional social science view of morality as convention about duty and obligation deriving from social custom. No wonder Westermarck came to be strongly resisted by both Freud and influential social scientists for most part of the 20th century (Wilson, 1998, pp. 194–5).

This is the first time that we learn from Wilson that Westermarck's view of morality was based on *moral emotions*. It is stated in just one sentence on page 195 in *Consilience*. But Westermarck's response to Frazer may not have been perceived as quite as simple and logical as Wilson believes. In fact, Westermarck's explanation remains puzzling to many even today. I will return to a discussion of what the Westermarckian view of the incest taboo may actually entail.

Can the dispute between Westermarck and Freud be settled empirically? Wilson believes it can and that the evidence leans strongly in Westermarck's favour. We learn again how it can be regarded as a series of multiple, successive barriers against incest:

> Up front is the Westermarck effect, the ancient sexual desensitization found in all other primates thus far and thus likely to be universal in humans. Next there is the dispersal of the young at sexual maturity, also a universal primate trait, manifested in humans by adolescent restlessness and the formal practices of exogamous marriage. The deeper psychological motivations of the dispersal behaviors and the epigenetic rules composing them remain unknown. Finally, there are the cultural incest taboos, which enhance the Westermarck effect and dispersal. *The taboos seem likely to have risen from the Westermarck effect but also, in a minority of societies, from a direct perception of the destructive effects of inbreeding.*
>
> (Wilson, 1998, pp. 195–6, italics added)

Wilson here approvingly cites Durham's comparative study of a sample of 60 societies (Durham, 1991), where he assessed to what extent its members had at least some awareness of the detrimental effects of close inbreeding. This was the case in about a third. Often explanatory folk theories or myths were developed (Wilson, 1998, p. 195).

Can we then say that by translating the Westermarck effect into incest taboos, humans have passed from pure instinct to pure rational choice, asks Wilson? This depends on what rational choice actually is, he continues. Here is how he believes it works:

> Rational choice is the casting about among alternative mental scenarios to hit upon the ones which, in a given context, satisfy the strongest epigenetic rules. It is these rules and this hierarchy of their relative strength by which human beings have successfully survived and reproduced for hundreds of millennia.

> *The incest avoidance case may illustrate the manner in which the coevolution of genes and culture has woven not just part but all of the rich fabric of human social behavior.*
>
> (Wilson, 1998, p. 196, italics added)

This is beautifully put, as it makes the Westermarck thesis the model for understanding the evolution of human behaviour. This rich statement might actually refer to several things. It seems to be emphasising the context dependence of epigenetic rules and the way in which their strength may vary. Might it even apply to circumstances where the necessities of human survival may have to go against the development or enforcement of incest taboos?

The Westermarck effect at the same time becomes a model for a naturalistic explanation of human behaviour. If only one type of explanation can be shown to exist – that of cause and effect – this would automatically preclude the appeal to purely cultural arguments of all kinds. For Wilson, being able to finally undermine the legitimacy of cultural explanations with their paralyzing effect on the social sciences is very satisfying (see his chapter on the social sciences in *Consilience*). Westermarck gets extra points for his early opposition to Freud, but it was only now that he could be vindicated.

The social conquest of earth: the Westermarck thesis as a keystone for Wilson's theory building

Wilson's popular book *The Social Conquest of Earth* (2012) is a compendium of many of his general ideas, this time addressing particularly the long process of human evolution and the role of gene–culture coevolution for the past and future success of our species. This is also a place where Wilson gives Westermarck and his thesis further recognition. After mentioning that some coevolutionary processes predate the emergence of *Homo sapiens* and even the human–chimpanzee split, Wilson states: 'One of the most important and best understood examples is incest avoidance'. He goes on to review the almost universal record of incest avoidance (with its known exceptions) and notes that in regard to brother–sister incest, a personal revulsion is reinforced by taboos and law. Here the content of the book becomes identical with *Consilience*. We encounter the three lines of resistance and again the tribute to Westermarck: 'This elemental response was discovered, not in monkeys and apes, but in human beings, by the Finnish anthropologist Edward A. Westermarck' (Wilson, 2012, p. 201).

Wilson also repeats his view that Wolf's results unequivocally favour the Westermarck hypothesis. He again describes the meticulously conducted study and its conclusion that the key inhibiting factor was close coexistence during the first 30 months of life of either or both of the partners and that the longer and closer the association had been during this critical period, the stronger was the later effect. The Shepher *kibbutz* studies, too, are cited as positive proof (Wilson, 2012, p. 202).

Then Wilson raises a new and provocative question. 'Is it possible that humans are *not* ruled by the Westermarck effect, but instead use their intelligence and memory to recognise that sibling and parent-offspring incest creates defective offspring?' (italics added). To this he answers no, referring to Durham's comparative study of 60 societies, of which only 20 showed any degree of awareness. (For Durham, on the other hand, the existence of incest themes in the myths of the great majority of the societies was taken as suggesting some kind of tacit knowledge.)[10] In this book we can also find the statement that was missing from Wilson's writings between 1980 and 1998: 'The Westermarck effect is an epigenetic rule of gene-culture coevolution' (p. 203).

Westermarck's own explanation of the origin of the incest taboo

Westermarck's own reasoning about the origin of the incest taboo is often seen as unsatisfactory. Many consider it unacceptable to proceed from the world of facts (the observation of sibling proximity) to the world of values (the creation of conscious prohibitions against sister–brother incest), or from the realm of biology to the realm of culture. For instance Durham (2005) writes about 'gaps' in the explanation of the taboo that invite 'filling', cites others who reason the same way and offers some solutions.

But is it possible that the 'gaps' are in the eyes of the beholder? It may be a matter of basic assumptions and research approach. It appears that few scholars have seriously tried to look at Westermarck's own thinking sympathetically 'from the inside', as it were, and approach his explanation on his own terms. Westermarck himself did not seem to think that his explanation of the incest taboo constituted a break with the other parts of his reasoning about incest avoidance; for him there was no 'gap' that needed bridging.

I believe that Arthur Wolf has put his finger exactly on the crucial matter and has captured Westermarck's thinking remarkably well. I will here follow him closely. Wolf thinks many scholars have been barking up the wrong tree:

> The solution to the problem of the incest taboo does not lie in discovering the hidden social or biological purpose it serves. *It lies instead in discovering how a behavioral tendency came to be the subject of a norm* [. . .]. Westermarck's solution to the problem – rarely mentioned by his many critics – is that the moral concepts which form the predicates of moral judgment 'are ultimately based on moral emotions'.
>
> (Wolf, 1993, p. 169, italics added)

In other words, Westermarck is interested in establishing continuity between individual emotions and social conventions. He distinguishes between two types of 'retributive emotions', positive or negative, acceptance or resentment, which consist of reactions to, respectively, the experience of pleasure or pain. The moral emotions are a special type of retributive emotions, which are distinguished by their 'disinterestedness, apparent impartiality, and flavour of generality' (Wolf, 1993, p. 169; Westermarck, ODMI II, p. 739). How do they come to have these qualities? Observe that Westermarck is looking for a particular *type* of answer to this question:

> The task is not, as Kant would have seen it, one of deciding when a judgment *is* disinterested, impartial, and general. It is instead one of explaining how some judgments *come to acquire an appearance of* disinterest, apparent impartiality, and the flavor of generality. To say that an act is good or bad is not, in Westermarck's view, to say that it is *intrinsically* good or bad; it is only, in Ginsberg's words, 'to refer the act to a class of acts *which have come to be called good or bad* because of their tendency to arouse these emotions'.
>
> (Wolf, 1993, p. 169; italics added)

So under what conditions do our emotional reactions feel and look disinterested, that is, not connected to our immediate interests? Westermarck singles out three conditions: when we feel sympathy, when we behave altruistically and when we feel disgust about something a person does, even if it has nothing to do with us personally. He calls this third feeling 'sentimental aversion' and suggests that we may be as upset with a person causing this feeling as if he had committed a real crime (Wolf p. 170, Westermarck, ODMI I, pp. 116–17). But how does an emotion take on the appearance of impartiality and generality? The answer is: when it is shared by the majority of the society and thereby takes on the status of a 'public emotion' (Westermarck, ODMI II, p. 740).

So, in other words, Westermarck is interested in how people's individual emotions can come together and give rise to shared, public emotions. He emphasises the fact that individuals are part of a community and that there are such things as public approval and disapproval. We can see how in his HHM book, he describes at length various public sentiments and the expressed rationales for the (changing) views of the appropriate punishment for different offenses. We should remember that Westermarck was primarily a sociologist and anthropologist and that he was interested in the very origins of public sentiments, customary practices and norms, which he saw as later developing into more specific prohibitions and laws. In other words, he was *empirically oriented*, not focused on abstract philosophical truths (although he saw his empirical study as a necessary basis for his later work on the moral emotions; Pipping, 1984). From the point of view of social psychology and nonverbal communication studies, his reasoning about a connection between individual emotions and public sentiment seems eminently plausible. Some of those sentiments may even have been contagious. 'Feeling each other's feelings' – including being indignant on behalf of others – was probably commonplace in early close-knit human communities.

And Westermarck does put his sociological foot down, as he declares:

> *Society is the birthplace of the moral consciousness.* The first moral judgments expressed not the private emotions of isolated individuals but emotions

> which were felt by the community at large. Public indignation is the prototype of moral disapproval and public approval the prototype of moral approbation.
> (Westermarck, ODMI II, p. 740; in Wolf, 1993, p. 170, italics added)

In other words, rather than invoking cognition or rationality, Westermarck deliberately aimed for a naturalistic view of morality as rooted in shared emotions. This is a highly unusual approach even today. Somebody who can be said to be on the same line as Westermarck, however, is E. O. Wilson.

Conclusion

The Westermarck thesis has played an important part in E. O. Wilson's theoretical development. It has been presented as an adaptation, an epigenetic rule, a tool for social engineering, a means for bridging the two cultures and even as an alternative approach to moral reasoning. Over time, it appears that the Westermarck thesis has become a keystone in Wilson's own creative theory construction. Currently the Westermarck thesis is getting increasing recognition. More evidence for it has been gathered, and various aspects of the thesis have been tested, among others, by evolutionary psychologists. The better the Westermarck hypothesis holds up, the more it will lend support to Wilson's own theorising (of course, there are also parts of his work that are quite unrelated to Westermarck). It is interesting to contemplate that as Wilson identifies more and more with Westermarck, he comes closer to a sociological point of view.

But how can we account for Wilson's strange seeming behaviour in regard to Westermarck over time? After figuring briefly in a footnote in *On Human Nature* (1978), Westermarck completely disappears from the radar, to reappear only 1998 in *Consilience* and then again later in 2012. I believe the explanation has to do with the sociobiology controversy and the general academic climate.

The Westermarck thesis played different roles in Wilson's different publications. In *On Human Nature*, its function was to bring in a recognizable human universal (the incest taboo) and thus a biologically grounded human nature at a time when the existence of a species-specific human nature was doubted. In *Genes, Mind, and Culture*, again, Wilson was striving to launch a brand new coevolutionary theory in the middle of ongoing controversy. In the expectation of criticism, a strong front was needed. Meanwhile Wilson and Lumsden felt they were in competition with other model builders, especially Cavalli-Sforza and Feldman (1981) (see Segerstrale, 2000).

The last thing in such a situation would be to weaken one's case by a careless move. I believe that Westermarck's name was deliberately omitted because of the potentially negative trigger effect of his name. This was a time of generally weak support for Westermarck (Allardt, 2000; Pipping, 1984), while there were hostile forces out there (Wilson had already had a pitcher of ice water poured onto his neck at the American Association for the Advancement of Science (AAAS) conference in 1978, and demonstrations abounded). And 'Westermarck' was certainly not a name to cite around social anthropologists or Freudians.

If this was the way that Wilson and Lumsden reasoned, they were right. The critics of sociobiology reacted predictably (e.g., Alper & Lange, 1981; Lewontin, 1981). Anthropologist Edmund Leach's *Nature* review (1981) was hostile. Interestingly, in a lengthy footnote in *Not in Our Genes* a trio of sociobiology critics soon voiced their skepticism of the incest avoidance thesis (without mentioning Westermarck), questioning the need for an incest taboo if the natural repugnance between siblings actually worked (Lewontin, Rose, & Kamin, 1984). We can guess that similar considerations suggested to Lumsden and Wilson to lay low also in regard to *Promethean Fire*. At the time, Shepher's *kibbutz* research had come under criticism, and this continued later (e.g., Hartung, 1985). And in 1986, the controversial matter of morality as a naturalist phenomenon certainly did not need the potential distraction of Westermarck's name.

So what had changed in 1998, when Wilson published *Consilience*? The Human Genome Project was under way, and 'genes' were receiving a more positive connotation. The culturist paradigm had for various reasons weakened (Segerstrale, 2000). But most importantly, in the meantime the authority of Freud, that cornerstone of culturism, had been seriously questioned by a number of scholars. He had not only been shown to be wrong, but also as having falsified his case studies of patients (e.g., Crews, 1996; Sulloway, 1979, 1991, 1992). The consensus in psychology was now that Freud was out (see Crews, 1996, for a number of references).

This meant that Westermarck was finally in – or at least that he was vindicated. Wilson could now feel free to mention Westermarck by name and give him due recognition while writing about 'Freud's fatal error' (1998, p. 81). In the meantime, the evidence for Westermarck's thesis had been further strengthened by Wolf and others. In conclusion, therefore, it looks as if Wilson deliberately kept Westermarck's name out of sight so that he would be able to work with Westermarck's important insights without further jeopardising his own ongoing theorising. Meanwhile he was waiting for a better academic climate to arrive, and this eventually happened.

Notes

1 Here we can see that Wilson, just like Westermarck, assumes that an instinct based on natural selection guides humans towards a 'gut feeling' of aversion.

2 Wilson provides a number of references to back up his statement, many of these from medicine. In later writings, he further updates the evidence. For a recent overview of the negative consequences of incest, see Lieberman and Antfolk (in press).

3 Wilson was keen on making social behaviour, too, part of the modern synthesis, whereby evolutionary processes were reformulated in the mathematical language of neo-Darwinian population genetics (see e.g., Segerstrale, 2002).

4 We see immediately that although the authors use the term 'coevolution', this is different from a coevolutionary model that lets biological and cultural processes evolve independently of each other.

5 This theme has sometimes been explored in novels (see e.g., Erickson, 1989). But there are also examples from reality of separated sisters and brothers who have fallen in love and even married (see Erickson, 2005). ('Natural experiments' of this kind are typically seen as indirect support for the Westermarck effect.)

6 They explain a cross-cultural diagram estimating brother–sister incest on page 135.

7 Wilson had in the meantime written a number of books on other topics, such as the need for saving biodiversity, and updated studies of ants, his favourite species. I am here tracing only his writings that are relevant for Westermarck.

8 Note Wilson's new specification of the sensitivity during the critical period: 'before one or the other is thirty months of age'. This new assessment corresponds to Wolf (1993) and (1995).

9 Pierre van den Berghe, however, explained this as rather representing a strategy for optimal inbreeding. He estimated that optimal inbreeding (whereby a human group would profit from increasing its inclusive fitness but still stay away from inbreeding depression) would be found with the marriage candidates having a relatedness lying between 1/8 (e.g., first cousins) and 1/4 (e.g., uncle–niece) (e.g., van den Berghe, 1983).

10 In *Consilience* it seemed that Wilson did in fact count with the possibility of observational knowledge. His pro-Westermarck resolve now appears to have hardened.

Part III

The emotional origins of morality

Sub-Editor: Camilla Kronqvist

9 The emotional origins of morality

Methodological issues

Camilla Kronqvist and Otto Pipatti

In the study of moral philosophy, it is usual to distinguish between normative ethics, descriptive ethics and meta-ethics. Among these, the first, rather schematically, is concerned with the formulation of rational principles for determining the right moral action. The second aims at descriptions of what different people believe to be morally right and wrong. It is often conducted in a comparative spirit, highlighting differences in moral beliefs. The third is concerned with clarifying the meaning or logic of moral sentences. It addresses questions such as, 'What does it mean to say that something is morally right?' or 'How does one determine the truth value of moral judgements?' Whereas anthropologists and psychologists have traditionally been the main contributors to descriptive ethics, the terms normative ethics and meta-ethics mark out philosophical debates.

In the present part of this volume we will not touch on the questions raised by normative ethics, except to voice Westermarck's skepticism towards attempts to construe ethics as a normative science. We will rather consider how Westermarck's reflections on the emotions as the primary source of morality anticipate a blurring of the lines between descriptive ethics and meta-ethics, a development that has only been strengthened within the emergent field of experimental moral psychology. Westermarck blurs the lines not the least because of his unique position as philosopher, sociologist and anthropologist. His works are a fruitful starting point for pursuing the intersection between these different fields. They also point to tensions that arise when different concerns and perspectives are brought together. Westermarck, for example, strongly endorses the view that 'The object of scientific ethics is not to lay down rules for human conduct, but to study the moral consciousness as a fact' (*ER* viii, cf *ER* 61 and *ODMI* I 18). What implications does such a view have for the study of morality and the different questions about normativity that can be raised in connection with it?

To provide a background against which the following three chapters can be read, we here give a brief overview of some significant moments in the origin of the moral philosophy turned into moral psychology. We also present some of the methodological key issues framing these debates as one way of approaching Westermarck's thought. These concern, on the one hand, the relation of both emotions and morality to what can be seen as facts and values, an objective and subjective reality. On the other hand, they raise more broad-ranging questions about the scope and methods of philosophy and its relation to empirical science.

From moral sense to moral sentiments: the early sentimentalists

The sentimentalist tradition in British moral philosophy originates in the early eighteenth-century theories of the moral sense, which challenged the Hobbesian view of human nature as fundamentally selfish and antisocial (Taylor, 2014; Turco, 2003). The concept of the moral sense was introduced in philosophical discourse by Lord Shaftesbury, who argued that human beings are capable of reflecting upon the affections that motivate them. It is by our natural and internal moral sense that we approve of the harmony of social and selfish affections in ourselves and dislike affections that are excessive or out of balance (Shaftesbury, 1999, pp. 172f, 196f.). Where Shaftesbury's language of the use of the word 'sense' was largely figurative, Francis Hutcheson developed the notion of moral sense into the moral sense theory proper. Hutcheson shifted the focus from the standpoint of the moral agent to that of the spectator, arguing that moral judgements of the actions of others are irreducible to self-interest (Raphael, 2007, p. 28). Because all virtue consists in benevolence, Hutcheson maintained that we only approve of actions motivated by benevolence, and we disapprove of actions that express other motives. Hutcheson attributed these disinterested feelings to the special moral sense, which he conceived as an independent faculty of mind, analogous with the external senses (Broadie, 2009, pp. 133–40).

In David Hume and Adam Smith, 'the moral sense' was altered into 'moral sentiments', conceived as the basis of moral judgement. They analysed the nature of these sentiments and explained them in terms of 'sympathy', or sharing the feelings of others. Hume claimed that moral approbation and disapprobation are special kinds of pleasure and pain, which are directed at the character traits of the person judged. Moral sentiments arise when we contemplate a person's character, not from the perspective of our own interests, but from a general point of view (Hume, 1739, 1751). Disinterestedness is a key element in moral approbation and disapprobation, and it is our sympathy with the feelings of those affected by the agent's character and action that makes vicarious and disinterested moral sentiments possible (Norton & Kuehn, 2006, pp. 965–7).

Westermarck, however, believed that Hume associated the moral sentiments too closely with pleasure and pain alone. In particular, Westermarck was profoundly influenced by Smith's theory of the moral sentiments. Smith ([1790] 1982) described moral approbation and disapprobation as kinds of vicarious gratitude and resentment – disinterested and impartial sentiments that we, as spectators, feel when we sympathise with the gratitude or resentment of others. Westermarck also sides with Smith in his rejection of Hume's view (stressed much more strongly in his *Enquiry*), that the moral sentiments are largely associated with our considerations of social utility (Lecture on Hume, manuscript, Westermarck 1913, pp. 82–3; Westermarck 1898, p. 306); Smith ([1790] 1982, pp. 89–90, 188). However, Westermarck's theory of the emotional origin of moral judgements is much more simple than Smith's. Smith postulates two distinct species of moral approbation and disapprobation. We feel approval when we sympathise with the emotion that motivates the agent, and we disapprove when we cannot share the emotion. In some circumstances, we also feel the additional desire to reward or punish the agent, and these moral sentiments arise out of our sympathy with the recipient's gratitude and resentment. Westermarck simplifies Smith's account by suggesting that moral approval and disapproval are retributive emotions akin to gratitude and resentment.

The logic of moral language: moral realism, emotivism and cognitivist expressivism

As part of his criticism against moral rationalism, Hume had argued that it is impossible to derive an 'ought' from an 'is', that is, a judgement of value from an observation of fact (Hume, 1739, pp. 468–70). At the beginning of the twentieth century, G.E. Moore, one of the founders of analytic philosophy, argued for a similar conclusion, although he resisted Hume's conclusion that moral judgements are an aspect of passion. He suggested that philosophers who attempt a reduction of the statement 'This is good' into statements such as 'This is pleasurable' or 'I feel approval towards it' commit a logical fallacy, the naturalistic fallacy. Rather, 'This is good' should be analysed as stating a moral fact, a fact of which we are intuitively aware. In this way, Moore appealed to a notion of ethical intuition. Nevertheless, he was not endorsing a moral sense theory, where ethical intuition is considered a discrete faculty (cf. MacIntyre, 2007, p. 15). His claim should rather be read as stating, without admittedly giving much argument for it, that the statement 'This is good' cannot be analysed any further. In that way his analysis is much more a remark about the character, form or logic of moral language than a comment on human nature, as in the case of Hume, Smith and Westermarck.

Nevertheless, Alasdair MacIntyre has argued that it was precisely this feature of Moore's logical analysis that created the ground for the emotivist arguments about the logical status of moral judgements that were to follow the moral realism inherent in Moore's thinking; 'Moore's followers had behaved as if their disagreements over what is good were being settled by an appeal to an objective and impersonal criterion; but in fact the stronger and psychologically more adroit will was prevailing' (MacIntyre, 2007, p. 17). The emotivists, some of them students of Moore, thus came to argue that moral judgements cannot be said to be true or false, but are expressive of emotions.

A.J. Ayer considered his form of logical positivism a continuation of 'the doctrines of Bertrand Russell and Wittgenstein', which he in turn saw as 'the logical outcome of the empiricism of Berkeley and David Hume' (Ayer, 1971 [1936], p. 9). He divided 'genuine propositions into two classes': the a priori and analytical propositions of logic and the hypothetical propositions of empirical facts (Ayer, 1971 [1936], 9). In doing so he leaned on Hume's distinction between 'relations of ideas' and 'matters of fact'. This left moral judgements as something less than 'genuine propositions', since they are neither analytic statements nor empirical hypotheses. Rather, Ayer proposed that they are expressive of emotion.

The words 'It is wrong', as in 'It is wrong for you to steal that', does not add anything proposition-like to the assertion 'You stole that', which, to Ayer, is an empirical fact. They only add a certain tone or emphasis to one's words (Ayer, 1971 [1936], 110). He does, however, add that they also serve as commands, with the intent of arousing emotions in others, so that the sentence 'You ought to tell the truth' also can be read as 'Tell the truth!' (Ayer, 1971 [1936], 113).

Ayer's *emotivism* and the related approach advanced in C.S. Stevenson's emotivism and R.M. Hare's prescriptivism about moral judgements are usually described as forms of *noncognitivism*. This is so because they deny that moral sentences can express something true or false. It is usual to distinguish these noncognitivist theories from more recent *expressivist* theories of morality, which agree that moral judgements are expressive of noncognitive attitudes, but are ready to admit truth-value to ethical propositions. Such views include Simon Blackburn's quasi-realism and Allan Gibbard's norm-expressivism.

The neo-sentimentalists

The early meta-ethicists were united in the conviction that, when asking questions such as, 'Is so–and-so good?' or 'Is this alternative better than that?' (Stevenson, 1937, p. 14), it is not clear to us what it is we ask for. The consequent task of clarification set off meta-ethics as a distinct field in moral philosophy, the task of which was to clarify the meaning of moral language (see also Joyce, 2008, p. 371). This notion of philosophy as doing preparatory, clarificatory, work, examining the kind of questions we have at hand, also marked off philosophy as distinct from the other sciences. Before doing empirical research or proposing normative principles for the right action, the task of philosophy, logic or conceptual analysis was to clarify the kinds of question empirical researchers could ask and what kinds of answer they could expect.

The ideas about what was involved in these kinds of analyses into the logic or meaning of language took many forms. One question concerned, and still concerns, the role philosophy should give to ordinary language, after Ludwig Wittgenstein, in his later thought, as well as other ordinary language philosophers such as J.L. Austin and Gilbert Ryle. They had suggested that attending to ordinary uses of language had a significant role in solving metaphysical riddles. This suggestion grew out of a firm belief that philosophy needed to stay free of metaphysical speculation, which in one way or other was regarded as meaningless.

In the middle of the century, however, there were growing attacks on this idea. This was partly due to two influential papers by V.W. Quine, which questioned the possibility of establishing logical truths through analysis of concepts (Quine, 1980) and suggested that philosophy 'must be methodologically continuous with science' (De Caro & Macarthur, 2010, p. 5, see also Quine, 1969, p. 126f). The growing dissatisfaction with conceptual analysis as the main philosophical tool led to a turn toward naturalism in philosophy. This development can also be seen in connection with a 'prevailing metaphysics' (McCann, 2006, p. 229) within analytic philosophy and the logical-positivist framework that from its beginnings had privileged empirical statements and the natural sciences in explicating meaningful language use.

This paved the way for a new wave of sentimentalists who sought to 'naturalise' ethics by relating morality to the emotions. The philosophers working within this field often take their inspiration from Hume and Smith and move within a growing field of moral psychology with a specific focus on experimental research (see, e.g., Haidt, 2001; Nichols, 2004; Prinz, 2007b). Within the philosophy of emotion, the neo-sentimentalists are usually critical of the cognitive theories of emotions that established themselves as the main philosophical interpretations of emotion during the 1980s and '90s (cf. Nussbaum, 2001; Solomon, 1980; Taylor, 1975). The cognitivists, among others, maintained that the distinction between reason and emotion was fraught with problems. They wanted to show that emotions have a rational or intentional structure by demonstrating that emotions rely on judgements and norms about what is reasonable and appropriate. With the new sentimentalists, however, came a greater push towards regarding both thinking and feeling as distinctive brain processes; cognition and affect.

Within moral philosophy, the methods employed by neo-sentimentalists are largely descriptive, recording what moral judgements people are inclined to pass on different scenarios. Rather than basing the philosophical argument on philosophers' intuitions about what we would say in certain circumstances, they want to begin with considering what people actually do say. The hope is to thereby be able 'to arrive at a more fundamental understanding of what guides people's thinking in the relevant domain' (Knobe & Nichols, 2008, p. 4f). They also suggest that the results of their experimental research can contribute to meta-ethical debates. There, they usually position themselves against noncognitivist accounts of moral judgements and propose a form of moral realism, which suggests that moral judgements refer to facts that can be true (cf. Prinz, 2007b, p. 167f and Chapter 12, this volume.) In this way they differ from Moore in suggesting that moral facts can be naturalised. They also testify to a desire to have their philosophical discussion describe features of the *world* rather than *language*.

Open questions and different perspectives

The questions raised here contribute to a range of philosophical debates. These deal with the following: (1) The role of emotion in establishing both facts and values, the subjectivity of emotion and their relation to an objective reality; (2) The status of the distinction between facts and values, objective and subjective in the first place; (3) The meaning of 'truth' or 'truths' in the context of emotions and moral judgements; (4) The kind of relativism we need to accept if we regard emotions as central to what more broadly can be called our moral understanding; and (5) How one should conceive the philosophical method and in what ways philosophy can contribute to our thinking about morality.

The following chapters articulate different ways of thinking about these questions. They ask how reading Westermarck, in the development of his thought and in relation to two of his critics and advocates, adds to our understanding of the relation between emotions and morality. The chapters on the relation of Westermarck's thought to that of G.E. Moore and Jesse Prinz also point to limitations of the empiricism that runs through these attempts to make sense of both emotions and morality. This is done by showing how the issues under discussion have been treated in competing traditions within philosophy.

In particular, the two latter chapters by Kronqvist lean on the philosophical tradition following the later thought of Wittgenstein. Within this tradition, there is agreement with the Humean idea that our thinking about morality, and thinking in the first place, does not rest on rational foundations. Yet, there is considerable discussion about how this bears out on how we should understand what is meant with speaking about 'emotions', 'nature', 'truth', 'morality', 'facts' and 'values' in the concrete contexts of our embodied lives as human beings. The fact that it has proven problematic to fit sentences such as 'Lying is wrong!' or 'This is good' into certain philosophical accounts of language use, it is suggested, exposes more problems with these accounts than with morality.

These views into the moral dimensions of our speaking about different aspects of human psychology, thinking, feeling, acting, seeing, were in part a response to Elisabeth Anscombe's call for a more 'adequate philosophy of psychology' before one could go on developing moral philosophy (Anscombe, 1981a [1958], p. 26). They provide yet a different philosophical way into thinking about our moral psychology. As an illustration of the kind of work philosophy, from this perspective, can do to elucidate the workings of language, we may consider Anscombe's own essay, 'On Brute Facts'. This is a short but incisive discussion of Hume's suggestion that we cannot derive 'ought' from 'is' (Anscombe, 1981b, see also Anscombe, 1981a), a notion that in different forms reappears through the previously mentioned discussions. According to Anscombe, Hume was right in thinking that we cannot derive an ought-statement from an is-statement. Nevertheless the form of a sentence alone does not reveal whether it is to be conceived as telling us something about 'what is' or about 'what ought to be done'. Therefore, any strict distinction between facts and values assumed to apply generally collapses. Rather than developing general theses about the nature of facts and values, reason and emotion, and basing one's account of morality on either, the task of philosophers is then to attend more carefully to the different work these distinctions do in different circumstances.

10 The evolution of Westermarck's theory of moral emotions

Otto Pipatti

Edward Westermarck's memoirs reveal that he had decided to write a work addressing the 'origin and development of the moral consciousness' already before his first magnum opus *The History of Human Marriage* was completed (MML, p. 100). He started working on the topic seriously during the first half of 1891.[1] From the very start, the gist of Westermarck's (HHM, p. 280) position was that '[o]ur notions of morality are closely connected with the instinctive feelings engraved in our nature'. However, it took some time before he was able to formulate a satisfactory theory of the nature of these emotions. Compared with a well-defined subject like marriage, Westermarck ponders on how 'morality extends into infinity'. Accordingly, 'It took a full five years before my ideas of the nature of the moral emotions had become more or less settled'. After that, it took ten more years before the disposition of the work was finally complete (MML, p. 102). This indicates a time frame of some fifteen years in the course of which Westermarck started seriously developing his own theory in 1892 and reached the phase of elaboration and literary composition during 1897 (ODMI I, p. 2; MML, p. 102). The first part of *The Origin and Development of the Moral Ideas*, where Westermarck presented his theoretical conclusions, was completed in late 1903 (MML, pp. 196–7), and the work was finally published in two volumes in 1906 and 1908.

The development of Westermarck's theory of moral emotions remains one of the many underexplored parts of his work. The present chapter explores his ethics by tracing and analysing the phases of its development, both its continuities and discontinuities. The three parts of the discussion proceed in chronological order. First, I discuss Darwin's account of the evolution of morality which marks the beginning of Westermarck's interest in the subject. Westermarck follows Darwin's theory of morality, structured around his account of the social instincts and the moral sense, at least from 1889 to the early years of the 1890s. Then, I look at the five-year formative period to which he points in his autobiography, when Westermarck departs from Darwin's moral sense theory and formulates his early theory of moral judgement by the early 1894. Lastly, during 1897 Westermarck frames the first outlines of his mature conceptions, finally published in 1906. After that, Westermarck's account of moral emotions occurs in nearly identical forms and even identical phrasings throughout his writings. In other respects, too, after the publication of the *Moral Ideas*, all the main features of his theory of morality remain unchanged.

Darwin leading the way

Westermarck's earliest contribution to debates on ethics is his lecture in 1889 on 'The theory of selection and its importance for the physical, mental and social sciences', delivered to the members of a student society (Nyländska Afdelningen at the University of Helsingfors/Helsinki). A year later he set forth his views in a lecture on 'Darwin's hypothesis concerning the origin of the moral sense' at the meeting of the Philosophical Society of Finland. Although the published minutes of the Philosophical Society on this topic are very brief and nonspecific (Westermarck, 1890, p. 160), the 1889 lecture (published in 1891) provides a useful but previously neglected source for exploring his early position.[2] Westermarck's nascent interest in the sources of morality was conspicuously inspired by Darwin's *Descent of Man* and his discussion of the moral sense. It is found, Westermarck submits, 'in one of the most ingenious chapters [Darwin] has ever written' (Westermarck, 1891b, p. 231).

In his earliest presentation, Westermarck begins by juxtaposing the moral condemnation of incest with how people generally judge acts as morally wrong. Whereas incest is considered immoral because 'such relationships contradict our natural instincts', Westermarck suggests that the same seems to apply also to 'other acts that we call immoral'. These actions contradict our social instincts, and thus the foundation of morality according to Darwin's evolutionary account (Westermarck, 1891b, pp. 231–2).

Darwin on social instincts and the moral sense

In *Descent of Man* Darwin attempted to demonstrate that the behavioural and mental differences between humans and animals are differences in degree rather than in kind. Humans and other animals possess 'similar passions, affections, and emotions' as well as similar mental abilities, 'though in very different degrees' (Darwin, 1981, pp. 48–9). These common features include the social instincts. Although Darwin shunned clear-cut definitions of social instincts, he listed four characteristic features. These are (1) the pleasure that social animals take in each other's company; (2) their tendency to feel sympathy, that is, the capacity to share and be affected by the feelings of others; (3) their altruistic disposition to perform various services for individuals of the same group; and, especially among human beings, (4) the strong desire to gain the approval or others and avoid their disapproval (Darwin, 1981, pp. 71–2, 86; Krebs, 2011, pp. 41–3). These four features are particularly bound up with sympathy, which Darwin regarded as the foundation stone of social instincts. For Westermarck, the specific importance of Darwin's account was that 'from these social instincts arise many of the actions we call moral' (Westermarck, 1891b, p. 232).

While arguing for the continuity between animal and human psychology, Darwin (1981, p. 70) emphasised that 'the moral sense or conscience' is the most important uniquely human feature. He conceived the moral sense as a natural extension of animal social instincts. Although Darwin and most of his later commentators often write vaguely about 'the moral sense or conscience', Darwin also distinguished between these two. He refers to 'the moral sense, which tells us what we ought to do' and, on the other hand, to 'the conscience which reproves us if we disobey it' (Darwin, 1981, p. 93). This distinction is endorsed by Westermarck's presentation of Darwin's theory, distinguishing between *remorse* and *moral sense*. According to Westermarck, 'remorse generally constitutes a criterion that shows that our moral sense has been offended' (Westermarck, 1891b, p. 233; see also 1890, p. 160).

Darwin (1981, p. 166) described the moral sense as 'the deep feeling of right and duty' which prompts altruistic acts. Young Westermarck was particularly impressed by Darwin's attempt to demonstrate 'how even the moral sense could be explained by the theory of selection' (Westermarck, 1891b, p. 231). Darwin assumed that moral behaviour, which is costly to the agent but beneficial to others, could arise because, in some special cases, natural selection acted on groups or communities rather than individuals. Thus, tribes consisting of individuals endowed with the readiness 'to give aid to each other and to sacrifice themselves for the common good, would be victorious over most other tribes; and this would be natural selection' (Darwin, 1981, p. 166) Later on, however, Westermarck provided a radically different description of the evolution of altruism.

The workings of conscience

Besides the idea of a moral sense, Westermarck (1890, p. 160) placed great value on Darwin's theory of conscience. Darwin proposed a four-part explanation for the evolution of conscience of which Westermarck highlights the first two, the social instincts and the increased cognitive capacities, which are associated with the ability to remember one's past actions and to reflect on them afterwards by comparing selfish actions with other-regarding alternatives. Westermarck quotes Darwin to sum up the question of the moral sense and conscience:

> Why should a man feel that he ought to obey one instinctive desire rather than another? Why does he bitterly regret if he has yielded to the strong sense of self-preservation, and has not risked his life to save that of a fellow-creature; or why does he regret having stolen food from severe hunger?
>
> (Darwin, 1981, p. 87; Westermarck, 1891b, p. 232)

The answer, Westermarck submits, is found in the conflict between different instincts and especially in the variations in their mental manifestation. What determines the workings of conscience is that the social instincts are more enduring than other kinds of instinct. Although the social instincts are 'ever present and persistent' in human nature, Darwin suggested that we are often moved by more selfish urges such as the instinct of self-preservation, lust and the desire for vengeance. These selfish impulses are transient by nature and, once satisfied, they lose their motivational power (Darwin, 1981, p. 89–90; Westermarck, 1891b, p. 232). However, as increased intelligence improves memory and imagination, images of past actions come constantly to the subject's mind. Because of this, Westermarck quotes Darwin's solution:

> [H]e will be compelled to compare the weaker impressions of, for instance, past hunger, or of vengeance satisfied or danger avoided at the cost of other men, with the instinct of sympathy and good-will to his fellows, which is still present and ever in some degree active in his mind. He will then feel in his imagination that a stronger instinct has yielded to one which now seems comparatively weak; and then that sense of dissatisfaction will inevitably be felt with which man is endowed, like every other animal, in order that his instincts may be obeyed.
>
> (Darwin, 1981, pp. 89–90)

Westermarck continues that because the individual now feels that she has neglected her social instincts in the past situation, the resulting feeling of dissatisfaction makes itself manifest in the form of regret or remorse. For young Westermarck, in sum, Darwin's account of the moral sense and conscience served as a clear and indisputable example of the necessity to incorporate the evolutionary perspective into psychological and ethical research (Westermarck, 1891b, p. 233).

Early theory of moral judgement 1894–1897

For all the impetus for the study of morality that Westermarck received from Darwin, he soon distanced himself from Darwin's reflections on the subject. During the first half of the 1890s, Westermarck was engaged in teaching and research at the University of Helsinki. In connection with this, he presented the outlines of his early theory of moral judgement. From this time, notes taken down by a student of Westermarck's (1894) lectures on 'Social Science' have survived, enabling us to explore how Westermarck's views take shape. Considering Westermarck's extensive contributions to moral psychology and the anthropological study of moral emotions, his sociology lectures are of special interest, for they reveal his first attempts to analyse moral emotions as the fundamental elements of human sociality. In 1895, Westermarck also presented his early theory in the Philosophical Society talk on 'Moral Feeling'. The most significant difference with regard to his initial position was that Westermarck no longer considered moral feeling to be the motive for altruistic action, but rather he saw it as the basis of moral *judgement*. As we shall see, his views on the subject remained the same until the spring of 1897.

During 1895 Westermarck also applied himself to the bodily expressions of emotions and lectured on the theory of emotion William James and Carl Lange had proposed independently ten years earlier. As Thomas Dixon points out, the traditional model assumed that the passions and emotions of the soul acted on the body and generated bodily changes. The revolutionary element of the James–Lange theory was to reverse the direction of causation and propose that 'the bodily changes that people had been inclined to call the "expression" of an emotion were in fact the primary constituent cause of the emotion' (Dixon, 2003, pp. 204–7). Accordingly, when we perceive a certain object or event, our body reflexively reacts in a certain manner, and emotions are the felt sensations of these physiological changes. Therefore, cognition does not cause emotions, but emotions are feelings of instinctive bodily response (James, 1884, pp. 189–90). Westermarck himself assigned bodily reactions a significant role in all mental activity, but added the caveat that there was not yet enough evidence to support the presumption of causality between bodily responses and emotions. 'I simply want to eliminate from the Lange–James hypothesis every reference to a causal relationship'. For Westermarck, affects and feelings are inextricably linked to sensations of bodily changes, but 'we do not know whether the organic sensations, that is, the sensations of the inner state of the body, exclusively determine the nature of the affects, we only know that they play an important part therein' (Westermarck, 1896, pp. 347–8, 353). Westermarck's position on this issue continues along similar lines in his mature theory of moral emotions.

In his early moral theory, Westermarck starts from the plain fact that we continuously evaluate the actions of others in moral terms. However, when we claim or think that certain acts are moral or immoral, Westermarck argues that we are not referring to the outward act itself, nor to its consequences, but to the intention behind the act. He defines intention as the will to reach a given end. What then makes us to regard some intentions as moral, or praiseworthy, and some as immoral, or blameworthy? Westermarck begins his answer by proposing that moral judgements are based on moral feelings which are of two kinds, moral approval and disapproval. Moral approval is a feeling of pleasure and moral disapproval is a feeling of pain (Westermarck, 1894, pp. 37–8).

The gist of Westermarck's position at this stage of its development is that moral approval and disapproval arise from our putting ourselves in the other person's place. When we observe the actions of others, we spontaneously form a 'vague idea' of how we would have acted if were in that person's shoes. He argues that the process is analogous to the kind of nausea we are prone to feel at looking at someone eating a substance we find revolting, or the dizziness we feel when seeing someone stand on the edge of a cliff (Westermarck, 1895, pp. 237–8). This particular psychological mechanism is the source of our fundamental feelings of moral approval and moral disapproval:

> The moral feeling arises when the intention of the person in action corresponds or fails to correspond with the intention I personally assume I would have if I were to be in his place. The correspondence between the ideas of intention arouses pleasure and evokes approval; the lack of correspondence evokes pain, disapproval.
>
> (Westermarck, 1894, p. 69; also 1895, p. 238)

Considering the interactive psychology through which the moral feelings arise, they belong to a wider class of feeling that Westermarck calls 'relational feelings'.[3] The relational character implies that *moral feelings arise out of the relationship between two ideas*. First, there is our interpretation of the agent's intention, and, second, our notion of how we would have acted in the same situation (Westermarck, 1894, pp. 33, 38–9). Westermarck also clarifies what he means by putting oneself in another person's shoes. It is not that I imagine myself to be the other person, but it is to imagine 'myself with the whole of my own psychological nature' in the same situation. Hence, the moral reactions carry a subjective stamp; they 'depend on the quality of my own will'. For example, when we feel that we would have resisted a certain temptation whereas the person we observe did not, we are roused to an unpleasant state of mind and we disapprove of her action (Westermarck, 1894, pp. 69–70).

At the same time, Westermarck emphasises that when we express moral judgements we are only concerned with how acts stand in relation to the person's will as a whole, that is, to her character. He makes the Humean point that humans have a natural tendency spontaneously to evaluate whether behaviours can be regarded as true expressions of character (Westermarck, 1894, p. 70). Unless we interpret human action this way, 'social life would be impossible' (Westermarck, 1892, p. 195). To employ Westermarck's example, if we consider a theft to be strongly influenced by some nonvolitional, external circumstances with uncontrollable strength to an average human being (severe hunger or a threat, for example), we are likely to think that the agent's action does not quite reflect her actual character. This, in turn, is likely to mitigate our moral disapproval. Most importantly, Westermarck already at this stage settled on his final position on moral judgement, emphasising that in our everyday reasoning, we in general do not reflect on how someone's character *has become* such as it is. We simply take one's character as it is, and as such it is the subject of moral judgement (Westermarck, 1894, p. 70; see also ODMI I, pp. 325–6). These discussions mark the beginnings of Westermarck's major interest in the psychology and anthropology of moral responsibility to which he devotes considerable amount of space both in the *Moral Ideas* and *Ethical Relativity*. Related to this, his view that the subject of moral judgement is not action, but rather intention behind the action, construed as an expression of person's character, remains constant in Westermarck's writings.

As in his earlier discussion of Darwin, Westermarck's early theory of moral judgement includes the fact that people have moral feelings also towards themselves. Instead of describing conscience in terms of a conflict between selfish and social instincts, he now describes its workings via the relational nature of moral feelings. We thus morally disapprove of our own action if we, afterwards, do not share our earlier intention (Westermarck, 1895, p. 238). Westermarck describes regret as a self-directed moral disapproval, which arises from the lack of correspondence of intentions. Correspondingly, if we assume that we still were to act in the same way if the situation should repeat itself, there is no conflict between intentions and our conscience is at ease. For Westermarck, these emotional expressions may be titled *conscience*, defined as 'the feeling of pleasure or pain which arises when the idea of one of my previous intentions comes into relation with the nature of my will as it is now' (Westermarck, 1894, pp. 39–40).

Moral feelings in social context (I): the rules of morality and the general will

Since the formulation of his early theory of moral judgement, Westermarck consistently looked beyond the strictly psychological facets of moral emotions as well as the dyadic relationship between spectator and agent. As a social scientist, he was equally interested in the social settings in which moral emotions are embedded, considering it crucial to study the influence of social context on these emotions. At the same time, he studied the manifestations of the same human emotions in different cultural settings, including their transmission in the form of moral customs and rules.

These efforts are clearly visible already in Westermarck's early work because relationality is not the only distinctive feature of moral approval and disapproval. The 1895 lecture makes clear that the second defining characteristic of our moral feelings is that the intentions we imagine ourselves to have in the agent's place must 'come within the sphere of recognised general rules, or the so-called moral commandments' (Westermarck, 1895, p. 239). In other words, our feelings of pleasure and pain are moral feelings only if our idea of how we would have acted in the same situation accords with some generally recognised moral rule.

One of the most neglected aspects of Westermarck's ethics concerns his major interest on the emergence and maintenance of moral norms. Westermarck's first analysis of moral rules is provided in his early lectures. The general moral prescriptions are expressions of what, at the time, he termed 'the general will' (Westermarck, 1894, pp. 29, 40) 'or the will that is dominating in a society' (Westermarck, 1895, p. 230). His early concept of the general will assumes that the 'character of people's will is more or less uniform' (Westermarck, 1894, p. 43; also 1895, p. 239). Because the moral feelings are relational feelings, differences with regard to them lead to conflict: 'If a person has a separate will which departs from the general will, he is disapproved of, as a result of the feeling of pain evoked by the lack of correspondence' (Westermarck, 1895, pp. 239–40). The resulting feeling is 'the public moral feeling of pain', and this moral disapproval, which is socially shared, gives rise to general moral rules such as the prohibition against killing and theft inside the community (Westermarck, 1894, p. 41).

Importantly, for Westermarck, general moral rules have originated spontaneously. Although some moral commandments have personal initiators, they 'have in general sprung directly from the general will, without having some special person announcing them or as an originator' (Westermarck, 1895, p. 239). In Westermarck's early theory, the fundamental moral rules found in all human societies express the similarity of human conation, that is, the fact that 'people are generally by nature equipped with social and egoistic instincts'. The Darwinian social instincts partly explain why human beings are generally reluctant to harm other members of the community and why they respond with moral disapproval when another person is hurt. In addition, Westermarck believed that we are led to the same direction by our egoistic instincts because we are likely to 'refrain from actions which may elicit revenge', that is, actions potentially harmful to ourselves (Westermarck, 1894, pp. 40–1).

Moral feelings in social context (II): the customs of society

In his discussion of Westermarck's early moral theory, Timothy Stroup emphasises that even though Westermarck's concept of the general will looks very Rousseauesque, he 'arrives at the notion as a Darwinian'. According to Westermarck, Stroup says, 'the concept is simply a (perhaps unfortunate) shorthand for the fact that the continuous operation of natural selection has resulted in a large measure of emotional agreement about important issues' (Stroup, 1982b, pp. 67–8). However, the Darwinian reading of Westermarck's notion of the general will – as simply the coincidence of individual wills – is not the whole story of the concept. If the rules of morality are manifestations of the general will, the concept must somehow take in also the factual diversity of moral norms. Taking this into account, already at this point Westermarck goes beyond the strictly Darwinian arguments discussed earlier. Westermarck's more inclusive view is represented in his account of the intimate connection between the general will and the customs of a society. For Westermarck, 'custom is always an expression of the general will materialised in action'. In his view, people are inclined to identify the rules of custom and moral standards, since customs are 'rules for conduct laid down by the general will' (Westermarck, 1894, pp. 41, 43). Referring to Westermarck's mature moral theory, G.H. von Wright (1982, p. 45) has rightly observed that 'the great emphasis Westermarck placed on the importance of custom [. . .] represents a continuation of his thinking from the time of the mid-1890s'. In his later work, Westermarck suggests that people tend to feel their emotion-based moral judgements as 'disinterested', 'impartial' and shared by others when they are in congruence with the prevailing customs, the social setting of morality.

When discussing the connection between customs and the general will in his early account, Westermarck first defines habit simply as something human individuals are accustomed to do. Custom, on the other hand, 'is a habit elevated to a norm of conduct'. Customs include a binding and regulating character, implying how one should act if one wishes to avoid social sanctions. Habits and customs are not the same, but in many cases customs are rooted in habits. He further notes that, for those habits that are for some reason, whatever it may be, regarded important for society, a general demand arises that they should be obeyed (Westermarck, 1894, pp. 27–30). The shift from habits to customs is further supported by the fact that action is commonly associated with normative expectations. 'In general one could say that moral rules are based on people's will: people act in a certain way and want that others should behave similarly' (Westermarck, 1895, p. 239).[4] He makes a similar distinction between habits and rules in his mature moral and social theory (ODMI I, p. 118; ER, p. 109). There the question of how public or 'social habits', due to the influence of collectively shared moral emotions, give rise to 'rules of custom, or institutions' (HHM I, pp. 69–71) becomes of even more pivotal importance. In this way, while retaining the idea of biological evolution as a central background element, Westermarck's theory of the development of morality assumes a social process of institutionalisation as a crucial condition for the development of moral norms.

The key shifts in 1897: the disinterested desire to reward and punish

As pointed out by von Wright (1982, p. 45), Westermarck's views developed rapidly and underwent great changes in the two years following the 1895 'Moral Feeling' lecture. This comcs up also in his correspondence where he writes, 'When I was teaching my little course on social science for the first time, I had a view on the moral emotion that I now regard fundamentally incorrect' (Westermarck, 1898). Broadly speaking, Westermarck arrived at the first outlines of his main conclusions, published finally ten years later in a more elaborated form in the first part of the *Moral Ideas*, by the spring of 1897 (von Wright, 1982, p. 45; Ihanus, 1999, pp. 257–9).

The first outlines of Westermarck's mature theory of moral emotions are set forth in his lecture 'The Predicate of Moral Judgements', delivered in late April 1897 to the Philosophical Society. Before turning to this, it is useful to consider his psychology lectures from the early 1897, which shed new light on this crucial period in the development of his thought. From these lectures, which have survived in the form of student notes, it becomes clear that at that time, Westermarck still adhered to the essentials of his early moral theory. However, it is his new account of the origin of moral rules that represents an important transition towards his mature theory of moral emotions. Instead of tracing the source of moral rules to the general will, Westermarck emphasises explicitly, as also later on, *the spectators' sympathetic resentment* as the foundation of moral norms (Hagelin, 1897, no pagination).

Westermarck suggests that moral rules are based on such approval and disapproval which concern not only one's own interests, but also actions directed towards others. He argues that human beings have a natural tendency to think that a person who intentionally inflicts harm on another and violates a moral rule should be punished. The most important alteration in Westermarck's argumentation is that this takes place because *such actions have a tendency to arouse the emotion of resentment or revenge*. This applies both to the person wronged and the spectators who are not personally involved. Drawing conspicuously on the main concepts of Adam Smith's moral philosophy, he emphasises that the rules of 'justice' are based upon 'the impartial spectator's sympathy' with the resentment of the injured party. Similarly, Westermarck is implicitly following Smith in suggesting that when we consider the agent as meritorious or worthy of reward, our moral judgement is based on our sympathy with the gratitude felt by the recipient of good deed (Hagelin, 1897).

Later in the spring of 1897, Westermarck presented his revised views to Philosophical Society. There he exposes the specific characteristics of moral emotions, 'so simple that they cannot be further analysed or described'. In his view, moral judgements are based on moral approval and moral disapproval, but contrary to his earlier position, these are not merely feelings of pleasure and pain. Instead they are *reactive emotions* directed towards other person as a response to her action. In Westermarck's words, 'moral disapproval is an emotion of indignation, in other words, an emotion that is accompanied by a distinctive desire to punish'. Respectively, 'moral approval is always accompanied by the desire to reward' (Westermarck, [1897] 1996, p. 300).

However, not all emotions of approval and disapproval are moral emotions. Here Westermarck introduces for the first time 'disinterestedness' as the descriptive characteristic by which moral emotions are distinguished from related emotions. This implies that 'the judgement "X is wrong" does not simply mean that the person making the judgement *desires* to reproach or punish [the agent], but that [the agent] *deserves* to be reproached or punished' (Westermarck, [1897] 1996, p. 300, my italics). Moreover, disinterestedness implies that we feel a desire to reward and punish also when we are not personally involved nor influenced by the action.

It was thus only during the spring of 1897 that Westermarck completely gave up his earlier twofold theory of moral feelings. First, he replaced the idea of relational feelings with the emerging theory of moral approval and disapproval as 'retributive emotions'. This was accomplished by combining his previous account of moral approval and disapproval on the one hand, and resentment and gratitude on the other, *into a single unified theory of moral emotions as retributive emotions*. By linking the moral emotions with the evolved tendencies for returning benefit for benefit and injury for injury, Westermarck gave the tendency towards *reciprocity* a central position. Second, he replaced his early ideas of moral commandments and the general will with the feature of 'disinterestedness', soon supplemented with the descriptions of 'relative impartiality' and 'flavour of generality'. These three descriptive characteristics, which found their way to the *Moral Ideas*, were presented for the first time in an article on the emotional basis of moral concepts published in *Mind* (Westermarck, 1900, pp. 184–5).

Simply put, this very heart of Westermarck's mature moral theory suggests that when the emotions of approval and disapproval are moral emotions, they are characterised by 'disinterestedness' – because they are felt as independent of any benefit to oneself; 'assumed impartiality' – because we feel that we do not favour any of the parties involved; and 'a flavour of generality' – because we assume that most other people in the society or social group would respond the same way in a similar situation (ODMI I, pp. 101–5). Due to these characteristics, moral approval and disapproval are typically felt as justified and generalisable: we feel that almost *anyone* who would happen to be in a similar situation would respond, or at least *ought* to respond, in the same way. These self-transcending features attach to moral emotions and moral judgements their specifically moral character, constantly cropping up in human interaction.

Neglected continuities

Despite the changes in his views, the main features of Westermarck's mature moral theory are not quite so late arrivals in his account of morality as previous commentators have assumed (Ihanus, 1999, pp. 257–9; Stroup, 1982b, pp. 67–8; von Wright, 1982, p. 47). Above all, the desire of retribution as a distinctive feature of human psychology and social reality figures already in his 1894 sociology lectures. In these lectures, Westermarck traces the occurrence of retribution in its many variations to the simple fact that human beings are on evolutionary grounds equipped with emotional tendencies which prompt the modes of behaviour we call revenge. In his definition, revenge, as an emotion, is the 'desire to inflict harm or pain to someone who have inflicted or wanted to inflict harm or pain to me'. Westermarck also observes that revenge in the simple form of direct and immediate retaliation is found in many animals. The feeling of revenge obviously does not always lead to the *act* of revenge, since 'this desire can in the blink of an eye be checked by other stronger emotions'. However, '*it may well be doubted that any human being would be totally free from this law of nature* [. . .] *that he would not feel the slightest desire for revenge, if another individual inflicted him some harm*' (Westermarck, 1894, pp. 46–8, my italics). Related to this, Westermarck (1894, pp. 46–56) discussed already at this point the origins of different forms of punishment in collectively shared punitive responses, thus anticipating his later and clearly more advanced contributions to the sociology, anthropology and philosophy of punishment.

The crucial difference between Westermarck's early and mature positions is that in his early account of retribution, he did not yet argue for the resemblance between the emotions of revenge and moral disapproval. As he would put it later on, 'It is the instinctive desire to inflict counter-pain – not necessarily physical – that gives moral indignation its most important characteristic' (ER, p. 85; also ODMI I, p. 92). Indeed, as Ihanus (1999, pp. 257–8) points out, some of Westermarck's letters suggest that it was especially due to his closer studies on the nature of revenge that gave him the keys to figure out the 'essence and basis of moral emotion' (Westermarck to Lagerborg, 2 October 1898, ÅAB: RL[5]). Similarly, as we saw earlier, it was not until the spring of 1897 that Westermarck was arriving at his equally central notion that the emotion of moral approval is akin to gratitude.

G.H. von Wright argues that '[i]t is not possible to say at what stage Westermarck came decisively under the influence of [Adam] Smith'. Referring specifically to Westermarck's account of moral emotions, he estimates that 'the main points of Westermarck's moral philosophy were arrived at without any direct influence from Smith' (von Wright, 1982, pp. 47–9). There is, however, more to add to von Wright's question. First, Smith's influence on Westermarck is apparent already in his early theory of moral judgement. His view that moral feelings arise from the process of placing oneself imaginatively in the shoes of the agent shows an obvious parallel with Smith's theory of moral judgement. More importantly, there are good reasons to claim that Westermarck reached his fundamental conclusions concerning the nature of moral emotions specifically as a result of studying Smith's *Theory of Moral Sentiments* more closely. As noted earlier, Westermarck's account of the spectator's sympathy with the resentment or gratitude felt by the person acted on, presented in his 1897 psychology lectures, simply repeats the key arguments of Smith's theory of moral judgement. Smith's decisive role in the formation of Westermarck's theory is also reflected in an application for a travel grant the same spring where Westermarck calls Smith the most important scholar so far to have addressed the origin and nature of morality.[6]

After this, Westermarck seems even to have understood his endeavours partly as a matter of putting Smith's theory of moral sentiments to an empirical test. Referring to his still unfinished *Moral Ideas*, Westermarck emphasises that the

> comprehensive study of the moral ideas of various nations and in various ages confirms the ingenious hypothesis set forth by Adam Smith, that resentment and gratitude belong to the root-principles of the moral consciousness.
> (Westermarck, 1900, p. 185)

It is noteworthy that Smith ([1790] 1982, pp. 67–78) examines moral judgements relating to whether a person deserves punishment or reward only in the second part of his work. For Smith, these judgements of 'merit' and 'demerit' are based on our sympathy or fellow-feeling with the gratitude or resentment felt by the person acted on. Because Smith's description of gratitude and resentment turned out to be crucial to the final turns of Westermarck's formative period, he might simply have studied the *Moral Sentiments* more carefully only during 1896, the results being visible in his lectures in the spring of 1897.

As to other continuities in Westermarck's thought, he pointed already in the same sociology lectures, albeit only in passing, to the characteristically impartial nature of moral feelings. Although an explicit account of impartiality would emerge in his moral theorising only several years later, Westermarck settled early on his final view that that revenge, properly speaking, is an emotion which more directly concerns one's private interests, whereas 'moral feeling is an impartial feeling'. However, Westermarck is not far removed from his mature position when he points out that we may likewise feel the desire for retribution when we, as spectators, only observe someone hurting or aiming to hurt someone else (Westermarck, 1894, pp. 46–8).

In his mature moral and social theory, Westermarck uses 'sympathy' to denote *sympathetic feelings or emotions*, that is, feelings and emotions that we feel as reactions to similar feelings and emotions in other people, arising through verbal, bodily and situational cues (ODMI I, pp. 109–10, 115–16; ER, pp. 95–7, 107). What follows is that besides responding morally to actions directed towards ourselves, we respond morally to how other people are treated. At the heart of Westermarck's scholarly interests lies the study of the classes of acts that cause us to feel hostility towards a person even though we are not personally at the receiving end of those acts. In this sense, the basic moral rules are spontaneous extensions of human emotional tendencies. Generally speaking, 'Customs and laws express the general feelings of the community and punish acts that shock them' (Westermarck, 1936, p. 236).

Conclusion

Westermarck arrived at his conclusions of the nature of moral emotions through two formative phases. During his preparatory work in the early 1890s he followed Darwin's moral sense theory without developing it further. Westermarck believed that certain behaviours are regarded as morally blameworthy because they conflict with our social instincts. He regarded the moral sense as a species of social instinct and as the motivational basis of actions that people regard as morally right. Like Darwin, Westermarck interpreted the moral sense and conscience as evolutionary products that function closely together. Remorse or regret expresses the feeling that we, by our own actions or omissions, have violated the moral sense, the motive feeling of right, wrong and duty.

Despite his initial enthusiasm, Westermarck soon rejected Darwin's view that moral judgements reflect agreement or conflict with social instincts. In his early theory of moral feelings, which he developed by 1894 and endorsed until the early 1897, Westermarck suggested that moral judgements are based on moral approval and disapproval, which he construed as feelings of pleasure and pain with a specifically relational character. The relational nature of moral approval and disapproval implies that they arise from the process of placing oneself into the shoes of the agent and the resulting agreement or lack of agreement between intentions. In addition, these feelings receive their moral character from their link with socially coercive moral rules and customs derived from the general will.

When shifting towards his mature theory of moral emotions in the spring of 1897, Westermarck again substantially altered his conceptions. What turned out to be crucial was Smith's account of the moral sentiments, even to the degree that Westermarck declared later that 'of all moral philosophers or moral psychologists there is none from whom I have learned anything like as much as from Adam Smith'.[7] Westermarck concluded that moral judgements are ultimately based on a specific pair of emotions seeking either to punish or to reward. These retributive emotions of approval and disapproval acquire their specific moral nature through the attributes of disinterestedness, assumed impartiality and from the fact that they are 'public emotions' common to a particular social group or society with regard to certain behaviours.

These major alterations notwithstanding, Westermarck's views exhibit various continuities as well. First of all, the constant elements concern his attempt to develop a theory of morality based on feelings or emotions. He was also early to analyse the connection between moral emotions and social customs and to develop an interest in the folk notions of moral responsibility. Above all, Westermarck considered the desire for retribution early on, both as a personal and vicarious response, as a tendency deeply rooted in human nature.

In conclusion, it should be stressed that in contrast to the pure emotivism of his early moral theory, in his mature account, Westermarck does not *identify* moral judgements with emotional expressions. Instead he observes that moral evaluation is often uttered 'without feeling any emotion at all ', or at least that emotions are involved to a very limited extent, insofar as a completely nonemotional mental state can exist at all (ER, p. 114; see also ODMI I, p. 4). Moreover, Westermarck does not claim moral judgements can in every individual case be directly traced to this pair of retributive emotions (ER, p. 62). Instead he wished to make a more wide-ranging and general point, namely, that 'without such emotions there would have been no moral predicates at all' (Westermarck, 1900, p. 184). Without the retributive nature of moral disapproval, in particular, 'moral condemnation and the ideas of right and wrong would never have come into existence' (ODMI I, p. 92; ER, p. 85). Westermarck sought behind moral judgements and norms emotions expressing 'a mental constitution which has been acquired through the influence of natural selection' (ER, p. 69). The retributive nature of moral emotions 'at once throws light on many of the otherwise dark corners of the moral consciousness'.[8]

Notes

1 Westermarck (1897) (Box 42a: Academica, work papers: 'To the University Senate, application for Hermann Rosenberg grant'), pp. 1, 4.
2 The English translation of Westermarck's lecture has been recently published (Westermarck, 2014). All Westermarck's writings prior to 1900 discussed in this chapter are in Swedish, and all translations are my own.
3 Westermarck's idea of relational feeling derives from the Scottish philosopher Alexander Bain (Westermarck, 1895, p. 237) and the Danish psychologist Alfred Lehmann (Ihanus, 1999, pp. 180–1).
4 In these transitions from habits to moral norms, Westermarck (1894, p. 27) places great emphasis also on supernatural beliefs which have 'always proved to be a prime mover in social life'. In his later works, too, Westermarck (ODMI II, p. 745; ER, p. 187) highlights 'the enormous influence' which the religious and supernatural beliefs have 'exercised upon the moral ideas of mankind'.
5 Available at: Ihanus, Lahtinen & Neuman (eds.), 2011. http://filosofia.fi/se/arkiv/text/3057 [Accessed 1 June 2015].
6 Westermarck (1897) (Box 42a: To the University Senate, application for Hermann Rosenberg grant), p. 2.
7 Westermarck (1914) (Box 78: Lecture in the history of philosophy, spring term 1914. Adam Smith), p. 21.
8 Westermarck (1913) (Box 78: Lectures 1913, Hume), p. 83.

11 Westermarck and Moore on the source of morality

Camilla Kronqvist

Edward Westermarck is bursting with enthusiasm when he describes his ongoing work on *Ethical Relativity* (ER), the book that sums up his main philosophical ideas. In a letter to his friend Rolf Lagerborg, he writes that his 'dearest book' . . . 'is at odds with almost everything written about Ethics in England during this century' (Westermarck to Lagerborg, 8 April 1928, ÅAB: EW). The task of this chapter, however, is not to deliver an assessment of how radical his thought actually was, but to spell out how Westermarck himself thought of this break with the prevailing tradition. By bringing emotion instead of reason to the forefront of moral judgement, he envisioned a kind of return to the British moral philosophers David Hume and Adam Smith. Hume, as many know, had famously argued that 'reason is, and ought only to be the slave of the passions' (Hume, 1739, p. 415).

The specific aim of this chapter is to clarify how some of the main points of Westermarck's observations in social anthropology and philosophy stand in the light of the criticism that British philosopher G.E. Moore levelled against them. The debate between them, although limited to a brief exchange, is concrete evidence of the relevance that moral philosophers of the time assigned to the themes that Westermarck was addressing. By linking the questions they raised to current debates on the relations between emotion and morality, I also show why these issues still pose pressing questions to moral philosophers of today. Attention to our natural, evolutionary history has become a prevalent element in the philosophy of emotion and in the explanations of morality. In this respect, Westermarck, to my mind, is really a pioneer, although his contribution has not generally been acknowledged.

The discussion proceeds in the following steps. In the first part, I present Westermarck's criticism of objectivist accounts of morality. This critique forms the background for his suggestion that morality should be seen as having its basis in our emotions and, in particular, in a form of other-directed retributive emotions. In the second part, I analyse the criticism that Moore directs at Westermarck, taking note of some crucial differences in their background presuppositions. In the third and final part, I argue that both parties share certain basic assumptions about the relation between emotions and facts; assumptions that are problematic. I articulate an alternative view of the role of emotion in moral understanding. This view incorporates what appears to be correct in the views of both Westermarck and Moore, but does not land in an unresolved conflict between subjectivity and objectivity.

Westermarck on moral emotions

The introduction to *Ethical Relativity* offers helpful suggestions for understanding how Westermarck regarded his contribution to moral philosophy. He polemically argues against the view that ethics is a '"normative" science, the object of which is to find and formulate moral principles and rules possessing objective validity' (ER, p. 3). He is right, one may feel, in writing,

> But while the evolution of other sciences has shown a tendency to increasing agreement on points of fundamental importance, the same can hardly be said to have been the case in the history of ethics, where the spirit of controversy has been much more conspicuous than the endeavour to add new truths to results already reached.
>
> (ER, pp. 3–4)

The numerous attempts to reach objectivity and lay down general moral principles, Westermarck claims, have not led to any consensus regarding the highest moral principle. He reviews several attempts at articulating principles that are objectively valid, among them different forms of utilitarianism, but he suggests that no philosopher has produced proof for the universal truth of their preferred principles (ER, pp. 4–14). He also rejects the notion that any moral principle could be intuitively evident, either to reason or to a special moral sense or faculty, considering the fact that numerous rival principles have been advanced as self-evident (ER, pp. 35–44).

To Westermarck, however, the divergence of views concerning the highest moral principle is merely theoretical. 'It has often been remarked that there is much greater agreement among moralists on the question of moral practice than on the question of theory' (ER, p. 45). He takes this as an indication that in our everyday moral life, we do not lean upon general moral principles. To capture the characteristic and poignant tone of his writings, a longer quote is in order:

> When they are trying to define the ultimate end of the right conduct or to find the essence of right and wrong, they give us the most contradictory definitions or explanations [. . .] But when they pass to a discussion of what is right and wrong in concrete cases, in the various circumstances of life, the disagreement is reduced to a surprising extent. They all tell us that we should be kind to our neighbour, that we should respect his life and property, that we should speak the truth, that we should live in monogamy and be faithful husbands or wives, that we should be sober and temperate, and so forth. This is what makes books on ethics, when they come to the particular rules of life, so exceedingly monotonous and dull; for even the most controversial and pugnacious theorist becomes then quite tame and commonplace. And the

> reason for this is that all ethical theories are as a matter of fact based on the morality of common sense.
>
> (ER, p. 47)

Having thus dismissed the normative attempt to formulate universal moral principles, Westermarck seeks to demonstrate the origins of morality in emotion. The main tenets of his moral theory had already been formulated in *The Origin and Development of the Moral Ideas* (ODMI). There he developed his thinking against the background of a large empirical material. *ER*, which is the focus of the present chapter, constitutes a more philosophical attempt to bring out his arguments and the implications of his ideas. It includes Westermarck's responses to critical points raised against *ODMI* (cf. ER, p. xviii) – including the criticism by G.E. Moore.

The moral emotions take centre stage in Westermarck's discussion. They 'belong to a wider class of emotions' called 'retributive emotions' (ER, p. 63) and divide into moral disapproval and moral approval. The first 'is a form of resentment' (ER, p. 89), 'a hostile attitude of mind towards a living being, or something taken as a living being, as a cause of pain' (ER, p. 68). The second 'is a form of retributive kindly emotion' (ER, p. 89), 'a friendly attitude towards such a being as a cause of pleasure' (ER, p. 86). The moral emotions are *retributive*, which means that they involve an impulse to 'pay back in kind', to inflict the cause of our pain or pleasure a corresponding feeling of pain or pleasure. More specifically, moral emotions are directed at actions 'assumed to be caused by' *the will* of another living being (ER, p. 172). What marks them off from the wider class of retributive emotions, which include vengeance and gratitude, is their *disinterestedness* and (relative) *impartiality* (ER, p. 108). The feeling of 'moral indignation', for instance, extends beyond the individual itself and applies to the subject's responses to actions by others when some other person than the subject is at the receiving end. As Otto Pipatti writes, 'we feel that almost *anyone* who would happen to be in a similar situation would respond, or at least *ought* to respond in the same way' (see Pipatti, Chapter 10, this volume, p. 124).

By seeking to ground moral judgements in emotions, Westermarck returns to a line of thinking advanced by 18th-century Enlightenment thinkers David Hume and Adam Smith. In Westermarck's writing there is a strong resonance of Hume's insistence that

> [t]he approbation or blame which then ensues, [when we look upon a crime as a crime] cannot be the work of the judgement, but of the heart; and is not a speculative proposition or affirmation, but an active feeling or sentiment.
>
> (Hume, 1751, p. 291)

It is, however, the moral psychology heralded by Smith that receives most favourable treatment by Westermarck. He claims that

> Smith's Theory of Moral Sentiments is the most important contribution to moral psychology made by any British thinker, and that it is so in the first

> place on account of the emphasis it lays on the retributive character of the moral emotions.
>
> (ER, p. 71)

This allegiance to Smith places Westermarck's thought in the Sentimentalist tradition. Nevertheless, his thought can also be read in relation to the meta-ethical debates that started to take form in the beginning of the 20th century. A central question here was how one should describe the meaning of moral language and, in particular, whether or not judgements such as 'It is wrong!' represent states of affairs that can be true or false. This discussion, as Timothy Stroup writes in his illuminative discussion of Westermarck's ethics, 'boils down to the question of whether we create or discover moral facts, whether we make them up ourselves or stumble across them' (Stroup, 1982b, p. 189).

Anticipating more recent discussions, Hume had denied the possibility of speaking of moral facts. Morality, to him, was a work of 'the human heart' or sentiment because, to him, 'Reason judges either of *matters of fact* or of *relations*' (Hume, 1751, p. 288, original emphasis), and the judgement, 'That is a crime', responds to no such a fact or relation (Hume, 1751, p. 291). It is, as Hume's claim is commonly described, impossible to infer judgements of value from judgements of facts, an 'ought' from an 'is' (cf. Hume, 1739, pp. 469–70). Westermarck follows Hume on this point. When I call an action bad, he argues, I am not expressing a moral fact about the action. '[T]he predicates of all moral judgments, all moral concepts, are ultimately based on emotion', he claims, and 'no objectivity can come from an emotion' (ER, p. 60). Certainly, it remains *a fact about me* that I feel this way (ER, p. 60), but this is a mere psychological fact about my moral consciousness. Therefore moral philosophy should only concern itself with studying such facts (cf. ER, p. 61). Westermarck, Stroup writes, even 'thinks that an account can be given, wholly in terms of human psychology and without reference to moral facts, of the way in which we come to use the moral concepts' (Stroup, 1982b, p. 189).

The common human tendency 'to assign objectivity to our subjective' moral experience (ER, p. 49), is thus also explained by reference to psychological and sociological facts about us. These psychological and sociological realities include the fact that agreement in moral matters simply tends to be greater than in ordinary matters of taste, and that 'generally speaking, people's inclination to generalise their judgments is greater in proportion as the impressions are found to be similar in each particular case' (ER, p. 50). For that reason Westermarck also emphasises the role of moral upbringing for impressing upon us the idea that some actions are right, whereas others are wrong, and 'the authority of public opinion, custom and law' which has prepared 'disagreeable consequences for those who act contrary to their decrees' (ER, pp. 50–1).

Moore on the meaning of 'good'

Westermarck portrays an objectivist moral philosophy he believes has lost its way. But what can the objectivists say in their defence? Here, G.E. Moore's criticism of Westermarck comes in as an interesting point of comparison. Active at about the same time as Westermarck, Moore (1873–1958) taught at the University of Cambridge from 1911 and was appointed Professor of Mental Philosophy and Logic in 1925, retiring in 1939. Together with Bertrand Russell, he is considered one of the founders of analytic philosophy as it developed in the beginning of the 20th century, partly as a reaction against the then prevailing neo-Hegelian British Idealism. His great work *Principia Ethica* was published in 1903,[1] and it was hailed by John Maynard Keynes, one of the members of the Bloomsbury group, as 'the beginning of a renaissance, the opening of a new heaven on a new earth' (MacIntyre, 2007, p. 14).

One of the main conclusions of that book is that 'goodness' is a 'simple notion' (Moore, 1903, p. 7) which cannot be further analysed. It is impossible to define the sentence, 'This is good' in terms of any other concept. Suppose, for instance, that a philosopher attempts to explain the meaning of 'goodness' in terms of pleasure. Moore then responds that although goodness may produce pleasure, it does not follow that what produces pleasure is also good (Moore, 1903, pp. 8–16). We can find pleasure in things that are not good in any moral sense; and we can imagine someone finding pleasure in evil. If we fail to honour the distinction between 'what is good' and 'what brings pleasure' we make ourselves guilty of the *naturalistic fallacy* (Moore, 1903, p. 13). In other words, we draw some conclusion about what is good and bad, that is, about a moral quality, from a psychological, and therefore natural, state: the state which consists in the feeling of pleasure. The fact that something is good or bad, however, cannot be established by the existence of particular emotional states. It can only be perceived through intuition. As mentioned, Westermarck is critical of the idea that the appeal to intuition could be used to justify any general moral principle. Moore's intuition, he remarks, tells him that it is impossible to define the good, yet this does not prevent utilitarian Henry Sidgwick from claiming that *his* intuition gives support to the opposite idea (ER, pp. 43–4).

Nonetheless, the naturalistic fallacy – the unwarranted transition from natural facts to moral facts – is a good place to start unravelling the critique that Moore directs at Westermarck in his 1922 article 'The Nature of Moral Philosophy'. There Moore accuses Westermarck of committing precisely this fallacy in his attempt to ground morality in emotions. He grants that Westermarck does not directly state that 'good' and 'evil' are simply names for whatever gives rise to moral approval or disapproval (Moore, 1922, p. 332). To Westermarck, it is not the case that *every time* I feel more indignant at one thing and less at another, I also assume that the one is more wrong than the other (cf. ER, p. 219). Nevertheless, Moore stills thinks that the word 'wrong', for Westermarck, involves reference to the feelings I *am inclined* to have. According to the position he attributes to Westermarck, my judgement that an action is wrong implies the statement 'that it is of a sort which *tends* to excite in me a peculiar kind of feeling' (Moore, 1922, p. 332).

Such a claim, Moore continues, is hardly plausible. If the only thing you and I do when we say, 'This is wrong', is to speak of our respective feelings, the utterance has a different meaning in your case and mine. Your 'wrong' points to your feeling, whereas my 'wrong' points to mine. If moral language only served the purpose or reporting such feelings, my saying 'this is wrong' and your saying 'this is not wrong' would be comparable with my saying 'I came from Cambridge today' and your saying 'I did not come from Cambridge today'. Because the two sentences do not concern the same thing, they do not contradict each other. If moral discourse were similar to this case, it would be impossible for us to *disagree* on moral matters. It would also mean that we are mistaken in believing that one moral judgement contradicts another. This, according to Moore, is a logical consequence of Westermarck's reasoning, but it cannot be right. Everything seems to suggest that moral disagreements exist, and this can only be the case if 'I mean by "wrong" the very same characteristic which you mean, and am thinking that the action possesses this characteristic while you think it does not' (Moore, 1922, p. 334).

Establishing the possibility of this kind of moral disagreement, Moore claims, is a presupposition of moral criticism. Although Westermarck suggests that some moral views 'mark a stage of higher refinement in the evolution of moral consciousness' (Moore, 1922, p. 334, quote from ODMI I, p. 89), Moore asks what, in Westermarck's view, it could mean to speak of one morality as higher than another. To Moore, the only thing Westermarck could mean by saying that A's morality is higher than B's, is that A's morality accords with his own, that it reflects his sentiments. Saying that A's morality is better than B's, however, Moore retorts, is to say that A's moral taste is better; which simply means that what A says is *true*, whereas what B says is false. In whatever way we try to make sense of this claim to truth, it is misleading to say that it is only a judgement about one's psychology (Moore, 1922, pp. 335–6).

To this, Westermarck responds that merely thinking that our moral judgements have a claim to truth does not disprove his view that the assumed objectivity of moral judgements is an illusion. Nevertheless, he claims that this 'denial of objective moral standards does not prevent my pronouncing moral judgements which are expressions of my own moral feelings' (ER, pp. 146–7). He further submits that embracing the subjectivity of moral judgements does not necessarily lead to a degradation of morality. It only provides another way of thinking of its basis:

> My moral judgements spring from my own moral consciousness; they judge of the conduct of other men not from their point of view but from mine, not in accordance with their feelings and opinions about right or wrong but according to my own. We approve and disapprove because we cannot do otherwise; our moral consciousness belongs to our mental constitution, which we cannot change as we please. Can we help feeling pain, when the fire burns us? Can we help sympathizing with our friends? Are these facts less necessary or less powerful in their consequences, because they fall within the subjective sphere of our experience? So also, why should the moral law command less obedience because it forms a part of ourselves?
>
> (ER, pp. 58–9)

Acknowledging subjectivity as an essential aspect of morality, and as an ineradicable part of ourselves, Westermarck contends, can even prove useful, because it can make people more tolerant and critical in their moral judgements (ER, p. 59). If we are able to trace 'the psychological and historical origin of people's moral opinions', it is possible to realise that many 'traditional notions as to what is good or bad, obligatory and indifferent' are based upon 'ignorance and superstition, or in sentimental likes or dislikes, to which a scrutinizing judge can attach little importance' (ER, p. 59). As he says, 'if it could be brought home to people that there is no absolute standard in morality they would perhaps be more apt to listen to the voice of reason' (ER, p. 147).

Summing up their arguments, Westermarck and Moore disagree on two accounts. First, they disagree on the nature, or correct analysis, of moral judgements, such as 'This is good'. In particular this disagreement concerns the predicate 'is good' and in what sense it can be true. Second, they disagree on the nature of moral disagreement. I will return to moral disagreement in the third section of this chapter. Before that, however, I will consider some differences in how they analyse the moral predicates.

As we have seen, Westermarck fully agrees with Moore that the sentence, 'This is right', cannot be equated with, 'I approve of this', in any straightforward sense. As Stroup argues, Westermarck is not an advocate of plain subjectivism (Stroup, 1981, pp. 217f; see also Kronqvist, 2014, p. 131–2). In this respect he also differs from the emotivist accounts of moral judgements that arose as a reaction to Moore's thought. Emotivists, such as A.J. Ayer and C.S. Stevenson, a student of Moore's, set out to show that the meaning of moral judgements was to express and arouse feelings or attitudes, and that they thus did not convey any moral facts. Ayer went as far as saying that 'one never really does dispute about questions of value' (Ayer, [1936] 1971, p. 69), claiming that all seemingly moral disputes were really disputes about empirical facts or logic.

Yet, it is noteworthy that Westermarck does not protest against the characterisation Moore gives of his analysis cited earlier. Westermarck submits that the only sense in which 'This is good' can be true, is 'if "good" expresses a tendency to feel moral approval', and if there really is such a tendency (ER, p. 142). In Westermarck's analysis, then, 'This is good', *pace* Moore, can be analysed further as, 'This has a tendency to evoke approval'. Nonetheless, to rephrase Moore's argument in terms of what has been called his 'open question argument': Can we conclude from the fact that 'It has a tendency to evoke approval' also that 'It is good' (cf. Moore, 1903, pp. 15–16)? Moore argues that we cannot. Whether the action or thing in question is good, one can say, is an open question. As for now, I also leave this question open.

The two philosophers also differ in the kind of analysis of the moral concepts and judgements they offer. Within the meta-ethical debates that Moore initiated, the focus is on the *meaning* and *justification* of moral language. Are they expressions of moral emotions, or are they descriptions of moral facts? Westermarck also shows an interest in the meaning of moral language and provides analyses of both moral concepts and the predicates and the subjects of moral judgements.

However, his explicit interest, also in these analyses, lies in investigating the *origin* of morality, and thereby in showing how moral judgements grow out of emotions and the customs of a society. As Stroup writes,

> Sometimes Westermarck does speak of moral concepts as 'direct expressions' of emotions, but more often he puts the relationship between a moral concept and an emotion at back stage: such concepts can be 'traced back to,' are 'ultimately based on' emotion; emotions are 'at the bottom' of the moral concepts.
>
> (Stroup, 1982b, p. 213)

Thus, whereas the meta-ethical debates strove to analyse the *concepts* or *logic* of morality, Westermarck's research was rather driven by a very strong *empirical* interest, bearing evidence of his work in sociology and anthropology. This difference in outlook is also reflected in their disagreement as to how 'good' should be analysed.

Moral disagreement

The issue of moral disagreement is central for both Westermarck and Moore. However, they disagree on what is entailed in the possibility to disagree. For Westermarck, the existence of apparently irresolvable philosophical disagreement as to what constitutes the basic principle of morality is one indication that morality is based on emotions, not on facts or principles. Moore's critique, on the other hand, presupposes that there can only be disagreement in so far as there is a fact to disagree about. Where both do agree, however, is in setting up an opposition between *what we subjectively feel* and *what is objectively the case* and regarding this as a sensible distinction. Their only disagreement is thus whether moral judgements are better to be seen as rooted in subjective emotions or as stating objective facts.

However, once Westermarck and Moore have made their choice according to this division, both seem to withdraw from saying that moral judgements are simply a matter of either-or. When Westermarck distinguishes moral approval and disapproval from other retributive emotions by emphasising their impartiality, he admits that 'if I pronounce an act by a friend or by an enemy good or bad, that implies that I assume it to be so independently of my friendly or hostile feelings towards the agent' (ER, p. 93). Yet because the moral emotions are emotions, speaking of 'objectivity', 'truth' or 'falsehood' in relation to them is impossible, except in the sense that it is true that one has an emotion, or a tendency to experience it (ER, p. 60). Moore, on his part, admits that claims about what is 'good' and 'bad' in some respects differ from other statements of facts. They are, for instance, not physically discernible. Yet, they are similar enough to statements of facts to admit of saying something about reality. They are predicates designed to mark out characteristics of actions and states of affairs, comparable to the colour yellow (Moore, 1903, pp. 7, 9–10).

As far as I can see, part of the trouble that both Westermarck and Moore encounter in fitting moral judgements into the categories of *either* subjective feelings *or* objective facts is connected with some central characteristics of emotion that are difficult to reconcile with this kind of categorisation. Despite their differences, they both accept that it is possible to give a solely psychological description of emotions. Thus they assume that the question of what someone *feels* can, in all situations, be answered without taking into consideration how things *are*. This assumption has been criticised by many more recent accounts of the rational and cognitive character of emotion. Approaching the question about how emotions relate to morality from such a perspective opens for new ways of exploring the kind of disagreement that is at hand when people disagree on moral matters.

One of Moore's aims in likening moral judgements to factual judgements is to show the sense in which moral judgements can be *true*. In that respect, they are similar to judgements about what someone can be said to 'know'. What it means to speak about 'knowing' cannot be regarded as a mere psychological question. Suppose someone says, 'The train is late' and the train is on schedule. Then we are justified in stating that the person does not *know* how things are. The correct expression is rather, 'She believes the train is late'. This distinction between a person knowing something and only believing it can only be made by establishing how things are. Is the train on schedule or not? The person's psychological state may be the same in both situations. In other words, if we say, 'She knows that the train is late', we, at the same time, *agree* that the statement 'The train is late' is true. This, we might say, is a conceptual remark about what it means to know something. Another way of putting it is that remarking upon the difference between knowledge and belief takes the form of logical clarification and not of psychological explanation.[2] According to Moore, a similar move can be made in the context of morality. 'Good' cannot be reduced to a matter of what someone wants, because a person's desires can be criticised in the light of what is good. When we turn to emotions, however, the same does not hold. 'The logic of emotion', we might say, lacks the central features that enable us to distinguish between 'wrong' and 'right'. We can agree that a person is angry, but deny that there is anything to be angry about.

In contemporary philosophy of emotion, this strict line between cognition and emotion has been questioned. Some evolutionary theories go as far as claiming that there is no real distinction to be made between cognition and emotion, regarding both as information-seeking processes (Cosmides & Tooby, 2000). Cognitive theories of emotion, for their part, have brought to the fore cognitive processes, such as thoughts and beliefs as significant parts of emotive processes, and suggested that expressing an emotion itself is to be regarded as a way of making a judgement (Nussbaum, 2001; Solomon, 1980). In some respects, Westermarck can be seen as precursor of these cognitive theories. This is, first of all, in the sense that to him, '[e]motions are determined by cognitions and differ in nature or strength according as the cognitions differ' (ER, p. 147). Second, it is because he suggests that 'subjective experience has been objectivised in the speech as a quality attributed to the object' (ER, pp. 143f, cf. also Stroup, 1981, p. 226). Third, one could argue that Westermarck's analysis of both the predicates and subjects of moral judgements involves an attempt to spell out the thought structure implicit in moral emotions, such as that they are directed at 'the person's will perceived as the cause of his conduct' (ER, p. 159). However, even though Westermarck regards emotion as something determined by cognition, he still retains a view of emotion as being distinct from cognition. Emotions change as cognitions change, but it is cognition that for Westermarck is the prime guide to the relevant facts of a situation. This still leaves out the sense in which apprehending the relevant facts of a situation rely upon emotion.

Within cognitive theories of emotion, the focus is rather on the rational, or intentional, structure of emotions themselves. Of central importance here is the *object* of emotion, that which the emotion is directed at or is about (see Kenny, 1963, for one early discussion of the formal object of emotions). Without attending to the object of emotion, they argue, it is impossible to identify *what* emotion we are dealing with in the first place. It is, in other words, the thought or the judgement that something is *unfair* that, in a specific situation, has us recognise certain reactions as *anger*. In another place (Kronqvist, 2005) I have commented on the risk of overintellectualising emotions by emphasising their cognitive character.[3] Nevertheless, these theories fill an important function in so far as they draw attention to the *internal* relation between attributing certain emotions to someone and describing a situation in a certain way.

Returning to one of Westermarck's examples, consider a person who feels moral indignation. Under what conditions will we be able to describe the person with these words? A central condition is that we must regard her emotive reaction as something that at least arguably is a reaction to a wrong, even if we may not agree with her that it *is* a wrong. It must seem to us that what has happened is plausibly understood as an injustice and that the person reacting in this way understands it in precisely this way. By calling her indignant and not, for instance, vengeful or aggressive, we imply that she has the capacity to understand her situation in the light of concepts such as 'justice' or 'being wronged'. In this sense, our description of the situation as 'unfair' is also an expression of our understanding of this as something that justifies being morally indignant. The situation gives the person *reason*, rather than *cause*, to feel indignant.[4] Furthermore, it gives us reason to understand the emotional reaction as indignant. By pointing to these aspects of the situation we make sense, or articulate, *what* someone feels, rather than present a cause as to *why* she feels the way she does.

Hence, Westermarck is right in thinking that it is a presupposition for mastering the concept of 'injustice' that one should spontaneously react to injustices with feelings of indignation. An important aspect of understanding someone's reaction as one of moral indignation is that we, too, may react with indignation at what happened. In this sense, certain spontaneous reactions are constitutive of the meaning of concepts such as 'justice'. Nonetheless, it is not possible to say what causes what. Does the situation give rise to the feeling, or the other way around? These two, I suggest, are involved in a constant dialectic. Sometimes, as David Cockburn notes, it is the expression on someone's face, such as a hurt expression, that makes us understand that something hurtful has happened. Other times, it is the kind of situation we are in, what I, for instance, know about the relationship of the hurt person to the one who hurt her that allows me to see her as someone hurt (Cockburn, 2009, p. 131). Thus, our understanding of what different situations and emotions mean grows in constant interaction.

We may of course agree with Moore that there is a difference between what we mean by *knowing* and what we mean by *feeling*. From the fact that a person has lost a loved one we cannot tell with certainty that he is grieving, even though he has reason for it and we are surprised if he is not grieving. Neither do we need to accept, from the fact that someone is angry, that an injustice has taken place, even if it is clear to us that the person thinks so herself. There is in that respect a greater indeterminacy surrounding our judgements regarding emotions than the ones regarding knowledge. Despite that, it is possible to criticise a person's emotions by appealing to aspects of the situation that have escaped him. The distinction between emotion and fact is of concrete use in particular situations where someone's emotion, for instance, is not in due proportion to the relevant facts. To be able to attend to the relevant facts in a moral light, however, demands an emotional engagement in these facts. To make someone see his situation differently from what he thought it was is also to make him feel differently about it.

The possibility of different perspectives on what the emotional meaning of a situation is, gives us a more nuanced and multifaceted picture of how emotions and facts are related. It also opens for new ways of understanding moral disagreement. Disagreement, for Moore, is primarily a matter of thinking that certain statements such as, 'Lying is wrong', are true or false. The realisation that different feelings are implied in different descriptions of a situation, such as, 'You lied to me!' raises instead the question what *moral meaning* it has to understand the situation in one way rather than another. What does it tell us about the particular individual that some concepts rather than others strike her as pertinent? What kind of criticism can we direct at her if we want to problematise her stance? Here there is reason to remark on at least two things.

First, the indeterminacy of many moral situations is an aspect of how individual concepts are to be applied to a specific case. Given that, say, 'That was a lie', already strikes us as the most appropriate way to describe someone's action, there is no further indeterminacy as to judging whether it was right or wrong. In other words, the judgement that lying is wrong (at least prima facie) is part of understanding what 'lying' means. Mastery of the concept of lying, thus, presupposes agreement of its moral status (cf. Hertzberg, 2013). Nevertheless, we may be in disagreement whether we are to call the particular thing someone said a lie. Did she lie on purpose, or did she withhold information in order not to hurt the other's feelings? The possibility of such disagreement grows as we make finer and finer distinctions in our descriptions of what somebody has done. Thereby moral discourse, *pace* Moore, is not so much a matter of determining whether a specific fact is good or bad, or whether a statement such as, 'Lying is wrong', is true or false. Many forms of moral disagreement rather revolve around the question of what descriptions apply to the situation we are in: 'Was it a lie or was it not?'

Second, as Westermarck recognises, the criticism we direct at a person's 'conduct and character' (ER, p. 148) goes beyond what we can see as their direct intentions or will. It also touches on their emotional responses (cf. Anscombe, 1981a; Cockburn, 1995; Cook, 1999). In what kind of spirit do people do things? What kinds of person do they become by doing and reacting in these ways? Is a person generous, considerate and loving in his meetings with close ones and strangers? Does he show reliability by always giving a helping hand, or can he be accused of being nonchalant or arrogant? Significantly, the form of criticism that we direct at someone's character in these cases is similar to the ones that we in other situations direct at single emotions. We can blame someone for reacting with envy to some happy news, but also for being an envious person.

It is easy to misunderstand this point if one thinks of a character trait or a personal quality as the *cause* of someone's envious acts and reactions. On this view, a person's envious character leads her to react with envy. We can dissolve this picture by looking again at how our descriptions of a person's emotions and character are conceptually connected. Remember that our typical ways of describing character traits proceed by pointing out single reactions in concrete cases, often in contrast with other possible reactions that we would have desired. In other words, an envious person is a person who tends to react with envy at the happiness of others and finds it difficult to rejoice in their happiness. Single reactions thus form a *pattern* in the life of a human being (cf. Wittgenstein, [1953] 2009, Part II, p. 174) so that we come to expect a self-aggrandising comment, a critical twitch of the lips when the successes of someone else is mentioned, from the envious person, we take his avoidance of certain topics for conversation as expressions of envy, and so on.

These remarks emphasise the need to pay heed to the role of emotion in how we morally understand things. Yet, they do not imply that we need to think of moral disagreement as a simple matter of taste. In this respect we can agree with both Westermarck and Moore that moral questions differ from aesthetic ones. Rather they are reminders of just how often we engage each other in conversations about the moral significance of how one does, and should, understand a situation. We express resentment and indignation when someone has acted callously, we rejoice at generosity, love and consideration, and we look to others for affirmation that ours is the way to feel. Westermarck is right that these conversations are rooted in, as well as expressive of, emotion. But just the same, we can claim that Moore is right in saying that they express a real desire to clarify how things stand. Especially, I would add, they express a need to find out how we, as human beings, stand in relation to one another.

This is also a point at which Westermarck's discussion does appear to have a considerable advantage over Moore's treatment of moral judgements. Westermarck realises that moral emotions are forms of our reactions to one another. Thus the moral emotions are characteristically set within an interpersonal context. This aspect of our moral conversations stands in need of more careful consideration. Such consideration may well lead us to recognise that the focus on *either* objective features of the world *or* subjective features of the selves is not a fruitful way to frame the problems involved in explicating what we do when we morally approve and disapprove of each other's thoughts, feelings and behaviour.

Conclusion

The purpose of this discussion was to bring out the disagreement between Westermarck and Moore on morality and the study of morality. By showing that emotion, or the moral emotions, can be a crucial starting point for criticising too sharp a division between subject and object, I have also brought out some assumptions they share about the nature of objectivity and subjectivity. I have suggested an alternative view on this as a way of dissolving the conflict between them.

Here I have focused on the interchange between Moore and Westermarck. Yet, the themes under discussion are closely related to many current debates that, contrary to what Moore believed, suggest that moral philosophy can 'be merely a department of Psychology' (Moore, 1922, p. 339). Now, we do not have to agree with Moore's version of moral realism, or with the kind of conceptual analysis he provides for the meaning of 'good', attempting to decompose sentences to bring out their smallest components. Nevertheless, we are well advised to distinguish between making conceptual remarks and providing empirical evidence when it comes to clarifying the role that the psychological study of emotions may have for the philosophical investigation of morality. This also goes for recognising the difference between speaking about causes and reasons in the case of emotion. Whereas the first are concerned with the *explanation* of behaviour, the latter rather concern the *justification* of emotion and action. Conflating the two easily leads one into Moore's naturalistic fallacy.

That said, we can grant Westermarck the notion that there are many important links between our understanding of emotions and our understanding of morality. What his discussion only partly captures, however, is the intimate relation between emotions and facts. Emotions, I have argued, cannot be distinguished from what we take as fact. In important ways, they set the stage for what counts as relevant facts in concrete situations.[5]

Notes

1 As a point of comparison, the first volume of *The Origin and Development of the Moral Ideas* was published in 1906 and *Ethical Relativity* in 1932.
2 Moore's 'naturalistic fallacy' may in that respect also be read as a form of philosophical criticism of 'psychologism'. Classical critics of the idea that one could understand the 'logical analysis of *the* mind' as a 'study of the minds of individuals' (Hutchinson, 2009, p. 74), include Frege (1893) and Husserl (1900). For a concise summary of this criticism see Kusch (2014).
3 See the following chapter for a neosentimentalist criticism of cognitivisim.
4 Compare this with Westermarck's claim that '[e]xternal events are generally the direct causes of our moral emotions' (ER, 149).
5 Work on this chapter was funded by the Kone Foundation. I want to thank the participants in the project 'Westermarck and Beyond: Evolutionary Approaches to Morality and Their Critics' for stimulating discussions and comments on this chapter. In particular Olli Lagerspetz has provided valuable comments and suggestions throughout the work on this chapter. Otto Pipatti gave me significant comments on one of the final drafts.

12 Emotions and moral relativism

Prinz and Westermarck

Camilla Kronqvist

One of the primary aims of Westermarck's two great works on morality and emotion (ODMI I-II, ER) was to disprove the idea that there could be universal principles for deciding what is morally wrong or right, available to us either by reasoning or by some form of intuition. To him, this idea failed to do justice to a sense in which moral predicates such as good or bad, just or unjust, do not describe properties of facts in our world. Instead, he argued that these concepts grow out of moral emotions, which are retributive in character. The consequence Westermarck draws from his idea that 'all moral concepts are ultimately based on emotions', is that there can be no moral objectivity, since 'no objectivity can come from an emotion' (Westermarck, ER, p. 60). This forms the background of his relativism. In the present chapter I consider this notion of ethical relativity in comparison with the arguments for moral relativism raised by neosentimentalist Jesse Prinz in his *The Emotional Construction of Morals* (2007b).

Prinz is interesting because he initially credits Westermarck with 'having recognised the link between sentimentalism and relativism' as well as the 'value of anthropology and history in investigating morals' (Prinz, 2007b, p. viii). Yet, he does not discuss Westermarck's moral theory in any more detail in the book. It is rather Hume's moral theory that receives more attention. Prinz's active endorsement of moral relativism, however, is an interesting point of departure for surveying the consequences of thinking of moral judgements as originating in emotions for the purported relative character of morality. Not least since moral relativism is a position much more widely adopted by anthropologists than by philosophers, partly because of a lingering belief within philosophy that endorsing relativism implicates one in Moore's naturalist fallacy (see, Kronqvist, Chapter 11, this volume, and Österman, 2014, pp. 97ff for more on this issue).

In the following discussion, I first examine the main tenets of Prinz's account of emotions and how it relates to morality.[1] I then survey the kinds of moral relativism suggested by Westermarck's and Prinz's theories and how they relate to their respective accounts of emotions, in search for differences and similarities that could help us determine whether an account of morality that gives central place to emotions necessarily leads to relativism. Inspired by John W. Cook's discussion of the role of moral relativism in anthropology and philosophy (1999), I then address the question whether it is possible to present empirical evidence for moral relativism and to what extent Westermarck and Prinz succeed in doing so. This raises some methodological questions concerning the suggestion, central to both Westermarck and Prinz, that the study of morality should describe the facts of our moral life.

Prinz's new sentimentalism

Jesse Prinz's account of the emotional construction of morals, as he calls it, is couched within a larger naturalist framework. The question that concerns him can thus be seen as locating 'the place of values in a world of facts' (Rorty, 2010, p. 57). There are four intellectual commitments in the naturalism that Prinz endorses. The first, *metaphysical naturalism*, includes the thought that 'our world is limited by the postulates and laws of the natural sciences' (Prinz, 2007b, p. 2). This leads to *explanatory naturalism*, according to which every explanatory model 'ultimately must be describable' in the language of a natural science (Prinz, 2007b, p. 2). The third, *methodological naturalism*, suggests that the methods used to 'investigate facts must be suitable to the investigation of natural facts' (Prinz, 2007b, p. 3). The fourth, *transformation naturalism*, embraces the thought that there is no position outside our present theories about the world from which to investigate it. Hence, all views about the world have to be changed from within (Prinz, 2007b, p. 3). These considerations usually lead naturalists to reject the idea of a distinctive philosophical method, such as conceptual analysis. Prinz, however, only considers it as an empirical method that is not 'especially powerful'. It is 'prone to error', because it 'proceeds through first-person access to psychological structures, or introspection' (Prinz, 2007b, p. 3).

It is against this background one can read Prinz's insistence that emotions are natural kinds, whereas morality is regarded as a cultural construction. It is also what leads him to give quite a narrow account of emotions and to favour causal and representational language in describing them. Emotions are to Prinz 'embodied appraisals'. They are 'caused by bodily changes' and they 'represent concerns' (Prinz, 2007b, p. 68). His account of morality is equally narrow. It is predominantly described as a culturally transmitted 'system of rules and values used to assess actions, agents and attitudes' (cf. Prinz, 2007b, p. 261).

Emotions, according to Prinz, belong to an affective system that is not mediated by the higher cognitive processes in the brain. Here, Prinz follows a line of thinking originally attributed to psychologists William James and Carl Lange, where emotions are regarded as somatic signals (Prinz, 2007b, p. 68; on Westermarck's views on James and Lange, cf. Pipatti, this volume). In assigning to emotions the capacity to represent, however, Prinz responds to the criticism that cognitive theories of emotion direct at the James–Lange theory. Cognitivists argue that 'thoughts', 'beliefs' or 'judgements' are necessary constituents of emotion. It is by virtue of these 'cognitions' that we distinguish between different emotions and speak of some emotions as warranted and unwarranted. For Prinz, this purpose is filled by a notion of representation that he takes from Fred Dretske's theory of meaning. Representation is here primarily seen as a matter of *detection*, where what is detected by emotions are things like losses for sadness, danger for fear or offenses for anger (Prinz, 2007b, pp. 61–5, 68). Significantly, such detection takes place without what Prinz takes cognition to be, that is, without the activation of the neocortex, supposedly the centre of higher cognitive functions (cf. LeDoux, 1996; Prinz, 2007b, p. 57). In this manner, Prinz believes that he has been able to repudiate cognitive theories of emotion.

Part of the disagreement with cognitivist accounts of emotions, however, appears to be at the level of definitions. Prinz ultimately equates emotions with somatic signals representative of a number of basic emotions or affects. In Paul Ekman's scheme, which Prinz suggests may demand elaboration, these include happiness, sadness, fear, anger, surprise and disgust. Cognitive theories of emotion rather term these states 'feelings', wishing to allow a richer concept for the word 'emotion'. In particular, they do this in order to accommodate the fact that these basic feelings are conceptualised differently in different contexts. It is, they argue, the different contexts and our means of articulating our concerns that provide us with the means of identifying various emotions. Prinz rather explains the range of emotions beyond the basic emotions and the cultural variation of emotion in terms of how basic emotions are either 'blended' or 'associated with mental files that calibrate them to concerns' (Prinz, 2007b, p. 67). 'The idea is that we have a universal emotional vocabulary, which is quickly co-opted by experience and tuned to culturally specific eliciting conditions' (Prinz, 2007b, p. 67). *Nostalgia* could, for instance, be seen as 'a blend of joy and sadness directed at the past' (Prinz, 2007b, p. 67), and jealousy as 'an acquired blend of anger, fear, sadness, and disgust' (Prinz, 2007b, p. 280, see also p. 67).

It is at this point that the moral emotions enter Prinz's theory. They are emotions calibrated to detect moral concerns: 'they promote or detect conduct that violates or conforms to a moral rule' (Prinz, 2007b, p. 68). He agrees with Hume that approbation and disapprobation are central to moral emotions, but proceeds to give a more elaborate list of basic negative and positive moral emotions (Prinz, 2007b, p. 69). His list is intentionally inconclusive, leaving us with the suggestion that it is an empirical question what is to count as a moral emotion. Major candidates at least include, on the negative side, contempt, moral anger and moral disgust, shame, guilt and the mixture of shame and guilt. None of these are to be seen as basic. Contempt is considered to be a blend of disgust and anger, whereas moral anger and moral disgust are particular forms of anger and disgust directed at transgressions against other persons or against nature. Prinz also introduces a distinction between reactive and reflective emotions. The former are directed at transgressions towards others, whereas the latter arise out of the experience of being a perpetrator (cf. also ER, pp. 94–5). An interesting category falls in between the two: reactions towards transgressions done to someone close to you by yourself or by others. The positive emotions of praise follow the axis of emotions directed at others and self, and include admiration, gratitude, gratification and dignity (Prinz, 2007b, pp. 79–82).

It is, however, in the context of discussing sympathy and empathy as positive moral emotions that the sentiments in Prinz's sentimentalism enter. They are defined as *dispositions* to experience particular emotions, where the notion of dispositions is also cashed out as 'real, physically implemented states of the mind' (Prinz, 2007b, p. 84); 'a psychological disposition is a standing state of the organism that can manifest itself as an occurrent state' (Prinz, 2007b, p. 84). In another place, he suggests that the sentiment, the standing disposition, can be called a *moral rule*, whereas particular manifestations of the sentiment, emotions, can be called *moral judgements* (Prinz, 2007b, p. 96).

Prinz's account is more elaborate and systematic than Westermarck's. Yet, it is also weighed down by the metaphysical baggage of Prinz's naturalism. Prinz does not propose that every explanation should be given in the terms of natural science. Rather he supports the idea that the social sciences, such as anthropology, document differences in the moral systems of various cultures on the basis of which philosophers can form ideas about, say, the legitimacy of speaking about different moral truths. Nevertheless, his desire to construe mental states as natural states leads him in a problematic way to posit a lot of things, such as emotions, moral concepts, moral rules, sentiments or dispositions, 'in the head' (Prinz, 2007b, pp. 91–7, cf. also pp. 57, 84; Prinz, 2007a, p. 272).[2]

I am sceptical of this suggestion, but not out of a wish to deny that something must happen in my head in order for me to be able to speak about feeling, say, jealousy. Yet, it is unclear what conclusions we can draw about what is happening in the brain, from the things we *say* we feel. Prinz's suggestion that jealousy is a 'blend of anger, sadness, fear, and disgust tuned to situations of suspected sexual infidelity' (Prinz, 2007b, p. 67) seems appropriate to the extent that we can elucidate what it means for someone to be 'jealous' by drawing on these other concepts. The 'jealous' person, as it were, is someone who may feel saddened and anxious at the thought of losing their partner and who may feel disgust at the thought of someone else touching their partner in a sexual way, or who easily gets angry when feeling neglected (see Ryle, 1955, for similar descriptions of the workings of the mind that do not rely on a stipulated inner machinery). In the light of this we may both regard 'jealousy' as a pattern of responses that finds expression in specific forms of anger, sadness, fear and disgust, but also as a fitting description of how someone reacts in a specific context. This, however, does not commit us to any specific view of particular occurrent or standing states of the brain. Rather, we can think of examples of how 'jealousy' conceptually ties in with other emotions as a matter of reflecting on how these words are used and how they figure in describing a person's relationship with the one they claim to love. Should we, for example, take their jealousy, if jealousy is the right word, to be an expression of their love, or does it rather reveal a failure to love, a lack of trust, a sense of insecurity or a need for control?

Whether 'jealousy' is a good description of what I feel, I argue, is not determined by the bodily (e.g., neurological) states I undergo. It is a question of in what ways I recognise myself in certain descriptions. What sense am I, or someone else, able to make of certain words in my life, given that we have learnt to use these words in appropriate ways? Do these words succeed in communicating what I feel? Am I true to myself and others about what it is I feel when saying I am jealous? Significantly, the difficulty of finding 'I feel jealous' a good description of myself, is not in the first instance a difficulty of identifying my emotion in the right way, but a difficulty of owning up to my emotions under this description. What does my jealousy make of me? What underlying but unarticulated problems with regard to the one I love do I need to confront? Do I have what it takes to change? Such reflections on how to see not only my emotions, but also myself aright can themselves be seen as involving me in a moral struggle.

I will follow up what is entailed in finding something a good description of one's emotion at the end of this chapter and ask what constitutes a good description of morality. First, however, I will turn to the different roles of 'nature' and 'culture' in Westermarck's and Prinz's varieties of ethical relativity.

From individual emotion to cultural societal rules

Westermarck and Prinz go some way towards uncovering the valuable lesson that our moral practices spring out of spontaneous, embodied responses endowed with meaning within a cultural context.[3] But how should we view this transition from the reactions of single individuals to public rules of behaviour? Moreover, how does this contribute to the relativist position?

For Westermarck, there is, strictly speaking, no real *transition* from the individual's moral emotions to those of society. The disinterestedness and impartiality that, to Westermarck, characterise the moral emotions already entail a social dimension, not least in the form of the individual's tendency to sympathise with the pains of pleasures of others. This kind of sympathy is not merely a matter of an association of ideas, of the kind that produces feelings of pleasure at seeing a happy face (ER, p. 96). Sympathy, when also concerned with how the other feels, 'requires the co-operation of the altruistic sentiment, or affection, a disposition of mind that is particularly apt to display itself as kindly emotion to other beings' (ER, p. 97). It is in virtue of being such creatures, living in groups, that mankind, according to Westermarck, has 'developed into the most gregarious of all animal species' (ER, p. 103).

This is part of the explanation Westermarck gives for the disinterestedness of emotion. To explain the claim to impartiality characteristic of moral judgements, Westermarck leans even more on society as 'the birth-place of the moral consciousness'. He writes, 'the first moral judgements expressed, not the private emotions of isolated individuals, but emotions felt by the society at large; [. . .] tribal custom was the earliest rule of duty' (ER, p. 109, cf. also ER, p. 50).

He goes on to elaborate on the role of custom in the establishment of moral rules: 'Customs are not merely public habits [. . .] but they are at the same time rules of conduct [. . .] And the rule of custom is conceived of as a moral rule, which decides what is right and wrong' (ER, p. 109). Westermarck thus explains moral rules by reference to customs, which in turn are explained by the public expression of emotion; the public sentiment that *anyone* would be apt to feel indignation at a given form of conduct. This gives Westermarck another means of arguing for the legitimacy of moral rules instead of invoking the notion of universal moral principles that apply to *everyone*.

In explaining the emergence of moral rules, and more particularly duties, Westermarck submits that moral disapproval must be seen as prior to moral approval (ER, p. 122). Moral disapproval serves to form prohibitions, 'the minimum of morality' (ER, p. 140), whereas the positive commandment to be the kind of person that merits moral approval or admiration only gains weight at a later stage in human moral history (ER, p. 171, cf. also pp. 134–8), 'the more scrutinizing the moral consciousness' (ER, p. 171) is. Moral approval, then, plays a significant part in the formation of moral ideals and aspirations (ER, p. 141). It is the basis of our concept of goodness, whereas the concepts 'duty', 'wrong' and 'right' grow out of public feelings of moral disapproval (ER, p. 122–8).

For Westermarck, then, ethics is relative or subjective in the sense that moral rules and judgements, and particularly the moral concepts, issue out of moral emotions. In his own words, attributions of the 'good' or 'bad' to character or conduct, spring out of

> generalizations derived from approval or disapproval felt with regard to certain modes of conduct. [. . .] [T]hey are tendencies to feel one or the other of these emotions interpreted as qualities, as dynamic tendencies, in the phenomena which gave rise to the emotion.
>
> (ER, p. 114)

There is, with regard to the moral emotions, a 'similar translation of emotional states into qualities' as in the case of calling something fearful because we fear it or admirable because we admire it (ER, p. 114).

In this respect, Westermarck and Prinz appear to be in agreement. Prinz construes being 'morally wrong/right' as a response dependent property and suggests that 'to harbor a moral belief is to have a sentiment of approbation or disapprobation' (Prinz, 2006, p. 36). Summing up the links that he sees between emotions and moral judgements, he writes, 'Emotions co-occur with moral judgments, influence moral judgments, are sufficient for moral judgements and are necessary for moral judgements, because moral judgements are constituted by emotional dispositions (either standing dispositions or manifest dispositions)' (Prinz, 2006, p. 36). As does Westermarck, Prinz takes this as grounds for rejecting the idea that morality could be objective.

Prinz, however, differs from Westermarck on one important point. In *ER*, Westermarck is adamant that there cannot be any moral truths because moral judgements depend on emotions. It can certainly be said to be true that someone has an emotion (ER, p. 60). He also concedes that 'the belief that gives rise to an emotion, the cognitive basis of it' can be said to be true or false (ER, p. 60). Yet, there is no other sense in which we can speak of one emotion as more true than another.[4] Prinz, however, argues for a form of moral realism where being wrong and right can indeed be perceived as moral facts and whereby a sentence such as 'stealing is wrong' can be true or false. It is *true* that 'stealing is wrong' if it is true that stealing causes us to feel 'angry, or guilty or disappointed' (Prinz, 2007b, p. 166). Moral facts of this kind are clearly nothing we discover, 'they are made by our sentiments' (Prinz, 2007b, p. 168). Once made, however, 'they can be perceived', they are no mere projections (Prinz, 2007b, p. 168). He admits, however, that on this scheme 'moral truth comes cheap' (Prinz, 2007b, pp. 292–3):

> There are probably thousands of moralities throughout the world, and an unbounded number of merely possible moralities. Each of these has a claim to truth. If moral claims are relativized to culturally inculcated systems of values in moral observers, then any claim about what morality demands can come out true.
>
> (Prinz, 2007b, p. 288)

For Prinz, moral claims are not only relative to the emotions of moral observers, but also to their 'culturally inculcated system of values' (Prinz, 2007b, p. 288). Here his account differs from Westermarck's in yet another significant respect, namely in the role accorded to nature and culture. Westermarck contends that moral judgements are ultimately based on emotions and that some forms of moral disagreement can only be explained as outcomes of differences in emotion. Nevertheless, his theory presupposes a great uniformity in our emotional reactions. Westermarck's aim with introducing examples of differences in moral views between cultures, and sometimes also in comparisons with animals, is often to argue for the universal prevalence of the retributive emotions that, for him, lie at the centre of moral life. Thus, although critical of universal *principles* guiding our moral judgements, he still operates with a notion of a universal *nature* belonging to us as living beings. He also points to the considerable agreement in moral *practice* that becomes evident when philosophers who defend very different moral principles spell out how we should live our life (cf. ER, p. 47, quoted in this volume on p. 130). In this emphasis on 'the comparatively uniform nature of the moral consciousness' (ER, p. 50), Westermarck is more Humean than Prinz is, for, as Prinz remarks, Hume's sentimentalism does not turn him into a relativist (Prinz, 2007b, p. 190).

Prinz, on the other hand, is critical of the idea of an innate human nature that shapes our moral values. Thus, he is also sceptical of attempts to ground morality in evolutionary biology or psychology, and implies that there is a danger of scientism in holding such views. '[M]oral rules', he claims, 'are not innate. Rather they emerge through interactions between biology and culture' (Prinz, 2007b, p. 274). In the bio-cultural account of morality that he suggests, 'the moral emotions have a non-moral foundation' (Prinz, 2007b, p. 104) because they are based on basic emotions such as anger, disgust and shame. Morality, or culturally distinct moralities, however, are on the whole, rather presented as cultural constructions, or, as in the earlier quote, 'culturally inculcated systems of values' (Prinz, 2007b, p. 288). The moral sentiments, or rules, form an important step between the biological and the cultural. They are constructed culturally, the 'mental files' that calibrate emotions, yet they can be localised in the head as standing dispositions.

Prinz's arguments for treating morality as a cultural construction partly rest on the ample empirical support he thinks descriptive moral relativism between cultures enjoys (Prinz, 2007b, p. 187–95). He also gives other reasons for considering 'morality as intersubjective' (Prinz, 2007b, p. 185). In particular, the semantics of 'ought' presuppose the agreement of a group; 'I cannot truly say that someone ought to do something without presupposing that this other person share my values' (Prinz, 2007b, p. 185). Further, 'If my values are informed by culture, promote culture and came into existence to serve cultural ends, then my values are very likely to be shared by others' (Prinz, 2007b, p. 185). Thus, he introduces the notion of 'communities of moralizers' (Prinz, 2007b, p. 186), which often do, but not necessarily, coincide with 'boundaries of groups which are pretheoretically regarded as cultures' (Prinz, 2007b, p. 186).

For Prinz, then, morality is relative on two accounts. It is relative to the emotional responses of an individual to what is wrong or right. It is also relative to the moral values of a community or a culture, and thus to the different pressures that local, material, geographical circumstances exert on the guiding ideas of a society. '[M]orality emerges as a system of rules for getting people to function collectively in stable and productive ways' (Prinz, 2007b, p. 185). In this way, Prinz's explanation of morality is functional in character, but he also introduces an 'evolutionary analogy' (Prinz, 2007b, p. 223) by regarding the cultural transmission of ideas as 'a function of fitness' (Prinz, 2007b, p. 220).[5] 'Cultures', he claims, 'consist in those things that are socially learned', psychological and behavioral traits such as 'habits, skills, ideas, values'. These are largely learnt through processes of conforming and conditioning (Prinz, 2007b, p. 184). Emotions are important among the factors conspiring to make a cultural idea, 'an evaluative belief', spread and be passed on, besides *material* and *narrative* factors (Prinz, 2007b, p. 220). The fact that '[e]motions increase attention and attention increases memory' (Prinz, 2007b, p. 221) turns emotional conditioning into an especially powerful means of learning.

Moral and cultural variability and the nature of normativity

Westermarck insists that irresolvable disagreements between moral judgements, and to some extent between moral theories, are due to differences in emotion between different individuals. Prinz rather submits that there is ample empirical support for a great cultural variation in moral systems. How should we understand this difference? Is Westermarck in some way blind to the evidence suggested by Prinz? Clearly, Westermarck registers much seeming disagreement and diversity in moral opinion, between individuals as well as cultures. These include differences due to insufficient knowledge of facts or insufficient reflection (cf. ER, pp. 147, 174). He also cites religious beliefs and superstition (ER, p. 187) as causes of cultural differences in attitudes towards human sacrifice (ER, pp. 187–9), suicide (pp. 189–92) and homosexual practices (pp. 192–6). A particularly interesting form of disagreement for the present discussion, however, are the differences in moral ideas that grow out of, what he calls, very different 'external conditions of life' (ER, p. 184). The examples under discussion include the killing or abandoning of elderly parents and infanticide. These examples are similar to the ones Prinz gives as reasons for accepting moral relativism; infanticide, Roman blood-sports, cannibalism and polygamy (Prinz, 2007b, pp. 187–95). The crucial moment in Westermarck's discussion, however, is that Westermarck does not regard the variability in beliefs encountered here as sufficient grounds for disproving the possibility of objective moral standards.

The idea that external conditions give rise to 'different ideas relating to the objective nature of similar modes of conduct and their consequences' (ER, p. 184), indicates that the disagreement in opinion goes beyond the factual beliefs that different agents hold as true. Such variation cannot be regarded as evidence for moral relativism, because it is not a case of two agents, placed in different times and cultures, who regard the *same* conduct in different kinds of light. Rather, their situations are objectively different to the extent that it is no longer possible to talk of the two agents as passing judgement on instances of the same conduct. Even if we admit that certain features of their situations are similar in some respect, objective descriptions of their circumstances make us realise that they are not the same (cf. Lagerspetz, Chapter 5, this volume).

John W. Cook speaks of the 'projection error' when he describes the effort by anthropologists and philosophers to prove moral relativism by citing evidence of different valuations of the same conduct (Cook, 1999, pp. 66, 69–70).[6] The aim of these anthropologists and philosophers is to describe a piece of conduct as relative to a specific cultural context, yet they still take themselves to recognise the same kind of conduct in another culture as in their own. Thus even though the purpose was to understand the other culture without appropriating it, the identification of a piece of behaviour as the same as in their own culture involves a projection of their own particular cultural understanding (Cook, 1999, p. 89–92). Michelle Moody-Adams offers a similar criticism of the attempt to empirically ground descriptive moral relativism, in that it does not attend carefully enough to the 'situational meaning' certain practices have in a given context. There can certainly be serious moral disagreement between cultures, just as there are serious moral disagreements within a culture, but Moody-Adams questions whether this is to be construed as a fundamental moral disagreement. She rather suggests that it is a disagreement about nonmoral facts; 'it would not be possible even to recognise some dispute as a *moral* disagreement if "ultimate" or "fundamental" moral disagreements were really occurring' (Moody-Adams, 1997, p. 16). This raises a basic methodological question about the role that empirical evidence can play in the study of moral psychology.

Prinz would certainly agree with Westermarck's statement that if the word 'ethics' 'is to be used as the name for a science, the object of that science can only be to study the moral consciousness as a fact' (ER, p. 61). Yet, despite some general similarities between Prinz and Westermarck's accounts, there are considerable differences in how they think that we should *take* the facts of morality on offer. In how far should we, for instance, see the moral lives of human beings as uniform, and in how far do the facts of these lives testify to a real cultural diversity in moral belief? In what respects can we speak of truth in relation to moral judgements? Finally, if it is possible for two philosophers to disagree on these questions, how do we determine which one gives us the best description of morality?

A first response is, of course, that the best description in some way needs to do justice to what morality is. A good description, one may think, is one that succeeds in describing morality. Therefore, we should first investigate morality to see which descriptions best capture it. Yet, it is advisable first to take a step back and consider what features characterise a *good description* of morality, and not straight away attempt to answer the question *what morality is*. This allows us to survey the kinds of *question* we are posing rather than the possible answers they can have. It also provides a way of scrutinising the tacit assumptions that guide us in our attempts to describe morality. We need not, for instance, accept the precondition that underlies Prinz's naturalism, *viz*. that any investigation has to be an empirical one. We can also seek to clarify what is meant by 'empirical evidence' and 'facts' in the first place. Asking what constitutes a good description of morality also draws attention to the evaluative, or normative, features of *giving a description* as such. This point is in part reminiscent of the Quinean claim that theories are always underdetermined by data, a point that Prinz would probably concede (cf. Moody-Adams, 1997, pp. 29f). Yet, it raises specific questions about the ways in which theories about what morality *is* can be challenged for failing to do justice to what is most central morally. Morality, Moody-Adams suggests, is a concept 'so complex that its content is "essentially contested," in W.B. Gallie's phrase, even by speakers of the same language' (Moody-Adams, 1997, p. 30, Gallie, 1956, see also MacIntyre, 1973). Asking what is required of a good description of morality, then, is itself a way of engaging in a moral inquiry as to how we should best understand morality.

Prinz is aware of the problems involved in painting too sharp a divide between fact and value when it comes to describing moral debates. In response to Moody-Adams's critique that the existing moral debates, which advocates of moral relativism cite in support of their view, should rather be perceived as indicative of differences in *nonmoral* beliefs, he remarks that, in some cases of moral debate, what the parties 'take to be constitutive' of the moral concepts under discussion 'is as much a matter of decision as discovery' (Prinz, 2007b, p. 192).[7] What concepts they use is itself an aspect of the parties' understanding, and not a ground for their view. Moody-Adams could thus be said to exaggerate the extent to which moral disagreement is based on differences in nonmoral beliefs, because the beliefs the parties hold are moral themselves.

Such an argument, however, can also be directed at Prinz's claims. Why should we, as moral inquirers, treat his description of morality as a system of values and rules, and the fact that there are crucial differences between these systems within different cultural groups, as a matter of discovery, and not as a matter of decision? If we think of it as a decision, then why should *his* decision to view morality in a particular way, or Westermarck's for that matter, carry any weight *for us*? Certainly some of their definitions could be of use in specific scientific studies, say, in neurology, psychology or anthropology. They could, for example, contribute to new ways of operationalising moral judgements or moral emotions and help to clarify what conclusions can be drawn from specific results. However, if the idea is that philosophers should make themselves useful to scientific research, would it not be more appropriate to address specific research designs, rather than proposing a general scientific theory of morality? If they, however, opt for the latter, there is also a need to clarify the concerns that motivate the search for such general theories. One apparent motive for both Westermarck and Prinz is to provide solutions to long-held philosophical disputes. Here, however, the value of such solutions also have to be judged against the general questions about morality to which these philosophers take themselves to be responding. What kinds of question are they asking, and what could count as an answer to them?

Another way of raising my concern about the general empirical setup of Prinz and Westermarck's writing is to ask what gives *authority* to their descriptions of morality. Appealing to empirical evidence of how people tend to react emotionally, or what they are inclined to say is 'right' and 'wrong' (cf. Prinz, 2007a, pp. 277–8 on semantic policies), is appropriate if used to support claims about how representative certain notions of what is morally right or wrong are in a given community. If, however, the concern is to determine whether a course of action is right or wrong in a concrete case, or to judge whether what people say is central to morality really succeeds in providing a good description of morality, appealing to such evidence of what people are inclined to think or say does not have any particular authority. My feelings may tell me to do one thing, even though I in other ways would concede that it is not right. Asking whether the majority of people would side with my feelings in deeming it right, also does not settle the question of what is right or wrong. The mere fact that many people would do as I do, if I am tempted to do something wrong, does not make the action right. In fact, it is possible to insist that something is morally wrong even though thinking so goes against common opinion.[8]

Such considerations play in the background for philosophers who search for universal principles that could *provide justification* for what we are to think of as morally right. Westermarck and Prinz are certainly right in criticising attempts to establish universal normative principles independently of our moral sensibilities. Nevertheless, it should be clear that my attempts to articulate *reasons* for taking something to be wrong, from *within* a certain vision of morality, is something distinct from trying to explain the origins of morality or the causes of my saying 'it is wrong'.

This also brings out problems involved in speaking about moral learning as a matter of mere cultural conditioning and conformity. A child comes to encompass moral rules partly through the prohibitions and emotional expressions of disapproval from its elders. This may make learning morality look like a form of cultural conditioning. Yet, it is central that the learning process does not stop there. As the child grows up, it also learns to use the moral vocabulary in new ways. It finds new meanings in these uses, meanings that others in turn may come to acknowledge. That is why we say that the child has *learnt* a moral lesson and not merely follows or reproduces a rule as a result of conditioning. As Cook remarks, 'Socially dictated attitudes and practices are not the stuff of morality; they are themselves among the things we subject to moral scrutiny' (Cook, 1999, p. 137).

It is perhaps on this individual level that Westermarck's talk of the moral consciousness as developing 'from unreflective to reflective' as moral emotions come to be 'more influenced by thought and reasoning' (ER, p. 147) is most poignant. This also goes for his suggestion that 'there are degrees of goodness and badness, virtues and merits may be greater or smaller, a duty may be more or less stringent' (ER, p. 218). Westermarck is reluctant to give much weight to the idea of one moral judgement as 'higher' than another, because that could only mean that the judgement is in accordance with one's own. Here, Prinz's submission that some moral values are better than others (Prinz, 2007b, p. 289), is more responsive to the recognition that some descriptions of the morally salient features of a situation stand out as more mature and insightful. Some views, as Prinz says, are 'inarticulate, or insincere' (Prinz, 2007b, p. 187) or vulnerable to self-deception (cf. Prinz, 2007b, p. 242). These views, rightly I would say, brings out a person's ability and preparedness to reflect on his or her actions and character as a central feature of morality. Yet I have difficulty seeing how empirical methods are expected to contribute to such moral reflection and to the deepened understanding of morality that such reflection may bring into being.

For Westermarck, moral progress is an outcome either of reflection, which often includes a reconsideration of the facts or the introduction of new facts that change our perception of the situation (ER, p. 147), or of a modification of the moral judgements by the sympathetic or altruistic sentiment (see e.g., ER, pp. 95f. and 207). The cure Prinz prescribes for moral change runs through the consideration of extra-moral standards or values, such as coherency and consistency (Prinz, 2007b, pp. 297–300), but also through genealogical critique (Prinz, 2007b, pp. 234–43). In criticising the idea of an external standpoint from which to scrutinise single judgements, both Westermarck and Prinz voice the idea that any criticism we can direct at our own moral judgements has to be internal (cf. Prinz, 2007b, p. 293). It has to start with our responses to what happens in our own life. I agree with them on that account. Yet, reaching for extra-moral standards for determining the moral value of certain ideas again relies too much on rational principles, in ways not far removed from Kantianism. Here, I see the kind of moral psychology that Iris Murdoch envisions in her works as a better means of improving our thinking about these matters.[9]

For Murdoch (1997a), the attempt to characterise human psychology is in itself a fundamentally moral endeavour, putting us in contact with a reality that is moral through and through (see also Kronqvist, 2009). Reaching moral insight, on this view, does not merely rely on the use of rational principles. It requires experience, effort and attention. This kind of venture, a transformation of our concepts, through personal experience, vision, inquiry and communication, she suggests, is a journey into a more personal realm, away from the public rule. It is a concern not only for what I should do, but for what I should be, constantly struggling in the direction of an 'ideal limit' (Murdoch, 1997a, p. 322). This view echoes the Westermarckian notion of the development of morality as a progression from prohibitions, expressive of moral disapproval, towards ideals of action and being that grant approval (ER, pp. 140–1). What Murdoch regards as central, however, is the change of personal moral vision, where one does not seek approval of one's actions, but a transformation of how one sees things in the light of the good.

Part of the reflection involved in moral inquiry or scrutiny involves giving reasons for or against alternative courses of actions. In that sense it has normative features in a way that the descriptive approaches to ethics and morality attempt to evade. Yet, a significant feature of moral reflection involves attending to the *descriptions* we use to characterise different possibilities in our moral lives. Moral questions do not start when we ask what to do – 'the chips are down', as Peter Winch writes, already in what we consider 'the alternatives' to be (Winch, 1972a, p. 178). By describing morality as a system of rules and values and not, say, as the contemplation of what is good (cf. Murdoch, 1997a, p. 311), we already commit ourselves to a *moral* view on what is central to moral life. Such pictures of ourselves and of our shared lives, images that surface in philosophical and scientific accounts of moral psychology, need to be scrutinised and not merely presented as a fact. We should ask how it matters morally to think of such pictures as true, and whether we really want to be like them. For, as Murdoch strikingly says, 'Man is a creature who makes pictures of himself and then comes to resemble the picture' (Murdoch, 1997b, p. 75). From such a perspective on what is involved in doing moral philosophy, clarity about what is entailed in speaking about both moral truths and the truth about morality, clearly does not come as cheap as Prinz suggests.

Conclusion

I have looked at some ways in which Westermarck and Prinz consider morality to be relative to emotions and cultural systems of belief. I have noted points of agreement in their discussions, but also moments at which their views diverge. The focus of my discussion, however, was not on the details of their accounts. Rather I used their disagreement on specific matters as a cue to raise questions about the empirical methods they both suggest should be applied in the study of ethics. A central problem in such investigations, I suggested, is that studying 'facts' does not, in itself, determine how we are to take those facts, for what purpose we study them and what concerns they are supposed to elucidate. What is more, there is a question about what we come to regard as a fact in the first place, and in what context it is a fact. For these reasons, I have suggested that no moral theory is purely descriptive. It always has normative features insofar as it presents a picture of what is central to morality. Such pictures can always be challenged in the light of the kind of vision they embody about what is central to *our* lives. In virtue of such considerations, we are not, for instance, compelled to accept the seemingly unproblematic assimilation of morality to a system of values and rules implied in Prinz's account, nor to accept as a fact that there is a diversity of such systems. I say this, not in defence of the search for one objective morality, but as a reminder of the fact that descriptions of moral psychology are not innocent. How we come to think of the differences between us, in what we view as central to morality and to different moral concepts, and what we recognise as significant similarities, is an issue that, in itself, poses us before moral questions that cannot be determined beforehand by a general theory.[10]

Notes

1 Readers who are not yet familiar with Westermarck's theory are referred back to the previous two chapters of the present volume for a more detailed account.

2 This is, in many ways, a consequence of Prinz's representationalist commitments. Huw Price gives a striking characterisation of two different ways of dealing with questions of language:

> Given a naturalistic conception of speakers, the addition of a representationalist conception of speech makes the object naturalist's ontological interpretation of the placement problem almost irresistible. Term by term, sentence by sentence, topic by topic, the representationalist's semantic ladder leads us from language to the world, from words to their worldly objects. Somehow, the resulting multiplicity of kinds of entities – values, modalities, meanings and the rest – needs to be accommodated within the natural realm.
>
> . . .
>
> Without a representationalist conception of the talk, however, the puzzle takes a very different form. It remains in the linguistic realm, a puzzle about a plurality of *ways of talking*, of forms of human linguistic behaviour.
>
> (Price, 2004, p. 87)

3 This is also a significant theme in Wittgenstein's later philosophy. See, e.g., Cockburn (2001), Hertzberg (2011), and Winch (1988) for explications of this idea.

4 In *ODMI*, however, we find a somewhat similar formulation to Prinz's. There Westermarck writes, 'it may be true or not that a given mode of conduct has a tendency to evoke in us moral indignation or moral approval. Hence a moral judgement is true or false according as its subject has or has not that tendency which the predicate attributes to it' (ODMI I, p. 17).

5 He does, however, suggest that the analogy between 'cultural elements' and 'genes' being transmitted, 'should not be taken too far' (Prinz, 2007b, p. 223).

6 Cook, who also discusses Westermarck's relativism, notes that Westermarck's discussion excludes these examples as proof of moral relativism (Cook, 1999, pp. 111–13). Cook fails, however, to notice that the reason Westermarck gives for accepting that there is genuine moral disagreement is difference in emotion, and not as Cook puts it, 'a clash of principles' (Cook, 1999, p. 114).

7 The specific example he discusses is the personhood of a foetus in debates about abortion.

8 This argument is, of course, no direct challenge to Westermarck's notion of impartiality as an almost ideal characteristic of moral emotions. This idea about what *anyone* would think is wrong or right is not, strictly speaking, something we find out through an empirical discovery.

9 See Broackes (2011) for a description of how Murdoch's thought also encompasses a kind of richer non-Humean naturalism. See also McDowell (1998) for another view of what naturalism in philosophy could entail, different from the Prinzian scientific one.

10 Work on this chapter was funded by the Kone Foundation and by Ella and Georg Ehrnrooth's Foundation. The presentation of Westermarck's moral relativism develops themes that I have previously discussed in Kronqvist (2014). My discussion has benefited from comments by Olli Lagerspetz, Natan Elgabsi and Otto Pipatti.

Part IV

Evolutionary psychology and morality

Critical perspectives

Sub-Editor: Ylva Gustafsson

13 Critical perspectives on evolutionary psychology and morality

Ylva Gustafsson

In its modern form, the claim that morality has a natural origin can be said to have had its starting point in Charles Darwin's *The Origin of Species*.[1] One of its implications was that not only human beings, but also animals, have a moral sensitivity and responsiveness to each other. In this sense, Darwin struck a forceful blow against the rationalist tradition of moral philosophy. His studies strongly suggested a naturalist and empiricist approach to questions concerning morality.

Darwin's evolutionary theory, in combination with his careful comparative studies of animals and human beings, had a considerable influence on Westermarck's reflections on ethics. However, Darwin was, of course, not the first to suggest that morality does not consist of purely rational reasoning, nor was he the first to argue for an empiricist perspective on morality. These ideas were already emphasised by David Hume and Adam Smith. What Darwin did was to claim that human beings are animals, not only in a physical sense, but also in a *moral* sense.

However, Westermarck was not only a follower of Smith, Hume and Darwin; he himself has had a considerable influence on evolutionary psychology.[2] Of central importance here are his writings on marriage, on the incest taboo and on the general evolution of morality. In this chapter we will, however, not consider Westermarck's work as such, but we will be addressing some central themes in contemporary evolutionary psychological theories on morality where Westermarck's thinking has had an influence. These areas are (1) the altruism/egoism debate, (2) rationalist and sentimentalist perspectives on morality and (3) contemporary normative and political perspectives on morality. In the following we will give a brief outline of these areas. In the last part of this chapter we will consider certain methodological questions concerning evolutionary psychological perspectives on ethics.

Altruism/egoism

The suggestion that man is essentially an egoistic being was forcefully expressed by Thomas Hobbes in *Leviathan* ([1651] 1985). He describes the natural state of man as a state of war of every man against every man. Society is created through a 'Covenant', or *social contract*, where men join together for mutual protection and cooperation. The fact that we are social beings thus depends, according to

Hobbes, on our capacity to form contractual relations and on the capacity to reason and consider such relations as beneficial.

In contrast to this, Hume and Smith argued that human beings have a natural tendency to feel *sympathy* for each other. In the sight of another person's suffering, I receive a similar, though weaker, impression of pain. This impression makes me inclined to help this person. Our social relationships, including our willingness to help, depend on such spontaneous emotional responsiveness.[3]

These two perspectives can be said to be forerunners of the debates on altruism and egoism. However, it was not until Darwin that the dichotomy between altruism and egoism could become a theme in an *evolutionary* approach to human psychology. Having originally concerned moral and political philosophy, the issues involved came to have quite a different character with the rise of evolutionary psychology. Of interest was not merely the behaviour of man, but the behaviour of man as well as other animals. With this also followed a change of vocabulary. Instead of talking about sympathy or compassion, the concept of 'altruism' became popular. Darwin did, however, not use the concept of altruism, but talked of 'sympathy' and 'the social instinct'. The term 'altruism' itself was coined by Auguste Comte. However, it was in the context of evolutionary psychology that it gained a huge influence. One reason for this is that 'older' concepts such as 'compassion' or 'sympathy' suggest human expressions of care and concern, and moreover, forms of care that often involve intellectual capacities such as thinking and speech. 'Compassion' and 'sympathy' seem to imply patterns of *human* rather than animal life. Therefore, if one wants to describe a pattern of care that is neither intellectual nor exclusively human, the concept of altruism appears more suitable and more neutral. This was a concept that Westermarck used. However, one may still ask whether a concept that connects human beings with other animals must *obviously* give us a more natural description of human life. Given this, why would the concept of altruism capture something more natural than is captured by the concept of compassion? These and related questions are considered in Chapter 14 by Ylva Gustafsson and in Chapter 15 by Hannes Nykänen.

The concept of altruism is, however, connected with another conceptual innovation that appeared roughly at the same time, namely the concept of 'empathy'. As already noted, Smith used the concepts of sympathy and compassion. However, the years when the theory of evolution was announced also saw the rise of psychology as a science. Adapting concepts that originally suggested moral evaluation into the new research context also meant their neutralisation, turning them into concepts that assumedly referred to mental states. Thus the concept of sympathy was replaced by that of empathy (by Edward Titchener, 1909).[4] Theories of altruism today commonly conclude that altruism depends on a cognitive mechanism of empathic imagination. Evolutionary perspectives on altruism and empathy are discussed by Gustafsson in Chapter 14.

The patterns of thinking described earlier and the corresponding changes in vocabulary can partly be seen in Westermarck's writings. However, Westermarck does not use the concept of empathy – rather, his key concept is that of 'retributive emotions'. One may say that Westermarck partly still belonged to the 'old school' of moral philosophy and partly he was influenced by the 'modern', scientifically oriented evolutionary trends of his time.

The concept of egoism, which is generally taken to be the opposite of altruism, has also assumed a different shape in evolutionary psychology. From having been a clearly critical concept, signifying blame, it has turned into a functional concept and into something good, something that enhances the survival of the species. The suggestion of the present chapter so far is that the altruism/egoism debate that has grown out of evolutionary psychology has radically reshaped moral concepts. Whether this is a good or a problematic thing we will not go into now.

These shifts in concepts do not merely reflect a change of theoretical perspective, but also one in methods of research. As already mentioned, there was a shift of focus from human to animal life; a shift that also meant a change in concepts. However, another shift is also visible *within* evolutionary psychology itself. The comparative method was, for Darwin and Westermarck, as well as other anthropologists of the evolutionist tradition, a central scientific method when doing research on emotions and on morality. However, in research today, the *experimental method* has become central. A question here is how to understand this shift. Has the shift towards experimental methods meant a more accurate scientific understanding of the natural aspects of morality? We will leave this question open for now.

Rationalism or sentimentalism

Another question relevant for evolutionary psychology is whether morality originates in reason alone or at least partly in emotions. As has already been noted, Westermarck, following Smith and Darwin, emphasised the emotional origins of morality. He was also strongly critical of rationalistic perspectives on emotion, such as that of Immanuel Kant. He was, however, not only critical of the claim that morality depends on rationality, but as Timothy Stroup points out (1982b, p. 148), he was also critical of the absolute conception of morality that Kant's rationalism implied. That is, he was critical of Kant's assumption that one can derive *general* rules of conduct deductively, without looking at how people actually live in various societies. Westermarck, in contrast, argued for a relativistic and empiricist perspective on morality. Morality takes different shapes depending on the various life circumstances that people live in. However, he also claimed that there is a natural and general structure in morality. This involves the retributive emotions, their impartiality and that these emotions are directed at persons and specifically at their will. In Chapter 15, Nykänen discusses the relation between Westermarck's and Kant's conceptions of morality. He argues that Westermarck's conception of disinterested retributive emotions is essentially Kantian, having connections especially with Kant's conception of the universality of reason.

However, again one can say that in contemporary evolutionary psychology there has been a clear shift in meaning concerning such words as 'reasoning', 'thinking' or 'rationality'. Rational thought, in both Kant's and Westermarck's lifetimes, meant reasoning according to general logical principles, as well as employing concepts that in themselves had a moral character. However, in contemporary evolutionary psychology, 'rationality' has come to mean an algorithmic function in the mind. Thus, the concept of rationality has changed: from something that implied the capacity to take responsibility and to make fair and true judgements it has been turned into a function of the mind that has no connection with such moral vocabulary. These changes perhaps indicate one reason why Westermarck's view on morality is rather seldom acknowledged in evolutionary psychological theory.[5]

Normative and political perspectives on morality

Even if Westermarck's writings to a large extent concerned questions about morality, they did not have a normative character. Westermarck was not interested in putting down rules for right or wrong conduct. Rather he described various patterns that human morality can have. As one can see from Westermarck's strongly critical stance towards Kant, he was sceptical of the idea of normative ethics as a science. However, a notable feature of Westermarck's influence on naturalist perspectives on morality today is that his thinking has come to be used for normative and even political purposes. This question is discussed in Antti Lepistö's chapter on Darwinian conservatives and Westermarck's ethics (Chapter 16). According to Lepistö, present-day conservative thinkers in the United States have been influenced by Westermarck's theories concerning maternal and paternal emotions, universal mating patterns and the incest taboo. These theories have been used to further conservative ideas concerning the 'traditional' two-parent family, the moral order of an autonomous civil society and natural law.

A question that arises here is whether these conservative perspectives ought to be seen as normative developments of Westermarck's descriptive ethics. That is, can one develop a normative morality from Westermarck's thinking? Or, more precisely, can one derive moral rules from the observation of facts? Or, in Humean vocabulary: can one derive an *ought* from *is*?

The answer to this last question has to be left open. However, Lepistö suggests that Westermarck was not a conservative thinker. Therefore, the contemporary conservative use of Westermarck's ideas may be a way of misrepresenting him. As Lepistö suggests, considering Westermarck's relativistic perspective on morality and his largely liberal attitude towards marriage, it is likely that he would have questioned many of the contemporary conservative interpretations of his own work, while being puzzled by the political outlook of some of his most dedicated followers.[6]

Some methodological issues and scientific language

As we have suggested earlier, Westermarck followed Darwin not only theoretically, but was also influenced by Darwin's careful comparative methods of research. As Timothy Stroup notes from Darwin's writings, 'Westermarck received a lesson in scientific method and the importance of fact as a control on hypothesis' (Stroup, 1982b, p. 37).

However, within a large part of evolutionary psychology there has been a clear change in the type of empirical methods mostly used. According to Stroup, 'Westermarck's procedure of drawing general conclusions from a mass of historically ordered data has been superseded by the experimental testing of physiological and behavioral hypotheses' (Stroup, 1982b, p. 158). One question that arises here is whether *experimental* methods can (or must) be considered more accurate, more precise, than the *comparative* method used by Westermarck. Has evolutionary psychology progressed towards more adequate empirical methods? In order to answer this question, it is important to reflect some more on the difference between the comparative and the experimental methods.

As noted earlier, with the comparative method, new ways of using moral concepts also were introduced. Thus, it involved not merely a new or broadened empirical perspective on morality, but it crucially changed certain concepts. The concept of altruism builds centrally on the idea that we can *compare* various animal species and their ways of interacting and caring for their kin. One problem, however, is that the concept is too generalising. The purpose with using it has been to describe similar patterns of behaviour in various animal species. Even though such comparisons can be illuminating, the comparisons risk also becoming too abstract. It becomes difficult to consider and give due weight to aspects that are central when we talk about human care and compassion, such as friendship, parenthood, having a sense of right and wrong conduct, etc. Bernard Williams (1985) distinguishes between 'thick' and 'thin' ethical notions. Among 'thick' concepts he includes 'treachery', 'promise', 'brutality' and 'courage', concepts 'which seem to express a union of facts and values' (Williams, 1985, p. 129). Descriptions in terms of these concepts tend to motivate actions. As an example of a 'thin' concept he gives 'the all-purpose descriptive term ought' (Williams, 1985, p. 130). One could consider the concept of 'altruism' as a 'thin' concept. Rather than singling out what is the truly natural feature of human care and compassion, the concept of altruism risks making our understanding of human life and what it means to care for each other too abstract.

Another concept that also often is used comparatively is the concept of emotion. For instance, various animals and human beings are often described as having certain emotions in common, and then these emotions are defined as 'basic' or 'natural'. However, one may ask what it actually means to define an emotion as basic in this sense. One problem here is that the claim that certain emotions are basic is often connected with the claim that these emotions actually have an instrumental evolutionary purpose. This is often connected with the tendency to present behaviouristic descriptions of these emotions. Such descriptions also then risk validating certain 'basic' emotional behaviours, whereas the absence of such 'basic' emotional behaviour is taken as a sign of abnormality.

To mention yet one concept frequently used in a comparative sense, consider the concept of 'maternal instincts'. It, too, relies on the idea that a natural aspect of parenting can be described and observed across different animal species. However, again, one may ask whether the concept might not actually obscure rather than throw light on the moral character of human parenthood. It contains the suggestion that parenthood mainly concerns physical care, whereas other aspects of the child–parent relation are obscured, such as the fact that we share a life history together, and a life where we *speak* with the child. Language and history are not integral to the concept of maternal instinct. Therefore, parenthood is reduced to something less than human. The suggestion that follows from this, then, is that a shared life history and language would not be intrinsic to what we mean by such words as 'parent' or 'child'.[7] This is a question discussed by Gustafsson in Chapter 14. Another reason why the concept of maternal instinct may be problematic is because it can easily imply a political agenda. Political aspects of considering maternal care as natural is a question that will be discussed by Antti Lepistö.

Having said all this, one may still ask whether it is not in some cases useful to make comparisons between animal species. Is it *always* wrong to claim that there is such a thing as altruism? Can we *never* talk of maternal instincts? Even though the concepts of altruism and maternal instinct may rely on problematic methodological assumptions, it is also important to note that the comparative method does not *necessarily* always have to be problematic. It depends very much on how the method is used and what claims are being made.

We shall now move on to consider the now so much more popular experimental method of research. How does this method differ from the comparative method? Has the shift from the comparative method to the experimental method meant a deepening of insights and more accurate empirical research practice?

Instead of observing an animal's behaviour in its natural life situations and making comparisons with human behaviour, as Darwin and Westermarck did, the experimental approach works by building up deliberately restricted and artificial experimental situations. The aim is to find systematic mechanisms of the human or animal mind. Especially within contemporary rationalistic theories of evolutionary psychology (often drawing on game theoretic considerations), such experimental methods have become very popular. What does this change of method then mean?

One way to answer this question can be to, once again, consider how moral language is changed by the use of the experimental methods. For instance, Cosmides & Tooby (2005) use the concept 'cheater detection module', referring to a postulated unconscious mechanism in the mind. However, such a concept has very little to do with what we ordinarily would call cheating. The presence of a 'module' is 'proved' on the basis of an experiment that has limited relevance outside the laboratory settings, such as the *Wason selection task* (Cosmides & Tooby, 2005, 2005). From these experiments, the discovery of an underlying mental module governing our moral conduct is announced. Here the moral word, 'cheat', has been turned into a scientific function (i.e., an algorithmic module in the mind enhancing the individual's fitness). A word that in our ordinary life suggests blame, such as the word 'cheating', is turned into a purely tactical concept. A similar change can be seen with 'egoism'. Ordinarily, the word 'egoism' is a word of condemnation. However, in evolutionary psychology it is turned into a morally neutral, genetic or mental mechanism. Consider, for instance, Richard Dawkins' 'selfish genes' (Dawkins, 2006). This shift is often reinforced by reductive experimental methods of research: the experimental method makes the shift of meaning from moral to scientific appear valid. Both comparative and experimental methods invest certain moral features in human life with new meanings.

The following three chapters address questions concerning the ethical implications of evolutionary psychological research on morality. Central questions in the following chapters are: What does it mean to make *empirical observations* of moral conduct? What *concepts* should we use when trying to understand moral conduct? Should we, or can we, use *scientific concepts* in order to describe the evolution of moral conduct? Can one derive *normative rules* of conduct from the observation of *facts*?

Notes

1 The claim that morality has a natural origin was expressed earlier by, among others, Jean-Jacques Rousseau. However, Darwin was the first to give this claim a scientific grounding.
2 The term 'evolutionary psychology' is often used in order to refer to a certain field of rationalistic theoretical perspectives on behaviour that arose especially from the 1990s onwards. However, in this chapter we will use this term more broadly to describe theoretical perspectives on the evolution of human morality from Darwin onwards.
3 Also Hobbes does talk about feelings such as 'Pitty', 'Compassion', and 'Fellow-feeling', which according to him 'ariseth from the imagination that the like calamity may befall himselfe' (Hobbes, [1651] 1985, p. 126 [p. 27 of the original edition]). However, Hobbes does not give this feeling any role in his explanation of the origins of morality.
4 Titchener was himself influenced by the psychologist Theodor Lipps' (1897) theory of '*Einfühlung*'. Lipps' theory of *Einfühlung* was originally concerned with aesthetics, discussing optical illusions.
5 Westermarck has, of course, been acknowledged as a precursor to evolutionary psychology, and then especially his writings on the incest taboo.
6 In *Three Essays on Sex and Marriage* (Westermarck, 1934b) Westermarck rejected Robert Briffault and F.W. Calverton's claim that he had written his book on the history of marriage in order to defend monogamy.
7 Soon after the publication of *The Origin and Development of the Moral Ideas I*, Westermarck's comparative method was strongly criticised by Émile Durkheim (see Lagerspetz, Chapter 2 of this volume). Durkheim argued that one ought to study specific societies and their life forms, because it is only as part of the specific societies that people's (or animal's) behaviour has meaning. Durkheim's criticism is important, although his functionalistic perspective on human social life also contains problems. We will not go further into discussing Durkheim in this introduction.

14 Why altruism may not be the right concept for understanding morality

Ylva Gustafsson[1]

Introduction

The question how to explain the fact that we are social beings has been a philosophical topic for a long time. If we are individual beings with individual feelings and thoughts, then why do we care for other people? Thomas Hobbes (1588–1679) suggested, on the whole, that man cares for others because those others are of use to him. Another explanation was put forward by Adam Smith (1723–1790) and later supported by Westermarck. Smith suggested that we care for others because we have a capacity to imagine ourselves in the other person's situation. This is the famous argument from analogy. Both these perspectives are still influential today in various fields, such as evolutionary psychology, developmental psychology, moral philosophy and economic theory. The present chapter aims to look at the role that these two perspectives have come to play in the thinking of some central modern evolutionary theorists' conception of altruism.

The modern evolutionary theorists I discuss are the primatologist and ethologist Frans de Waal, the psychologist Leda Cosmides and the anthropologist John Tooby. These represent two different perspectives on altruism. De Waal's perspective on altruism can be traced back to Adam Smith, Charles Darwin and Edward Westermarck. According to them, human beings have a natural tendency to respond emotionally to others, to sympathise or to feel empathy for others. Cosmides and Tooby have, on the other hand, a rationalistic perspective on altruism. According to them, altruism consists of an algorithmic reasoning mechanism, by which we calculate how to act in order to maximise fitness when dealing with others. Cosmides and Tooby's rationalistic perspective can be traced back to Hobbes, William D. Hamilton and John Maynard Smith.

It will be suggested in this chapter that both of the perspectives on altruism just described are problematic in their own ways. The main problems about them are, on the one hand, reliance on a transactional model of interpersonal relationships and, on the other hand, the view on interpersonal understanding as consisting mainly of the capacity for analogical imagination. I also claim that the two perspectives on altruism are, in some important ways, not actually opposites but share an instrumental conception about the character of human life. My suggestion will also be that the very concept of altruism in some ways perpetuates these assumptions and hence may obscure more than clarify the meaning of care and compassion in human life.

Although both perspectives on altruism often offer insightful observations of human life, their theoretical explanations may still be misguided. I suggest that some of their descriptions are too restricted in focus, too one-sided or too generalising. By bringing in alternative examples and alternative descriptions of the behaviour that the authors in question cite for support, I will point at other ways of describing human life. Still, even if I question these two perspectives on altruism, I also think that especially the 'theorists of empathy', that is, Smith, Darwin, Westermarck and de Waal, offer many sensitive reflections on the natural character of our shared human life.

Altruism as spontaneous emotional responsiveness

In *The Age of Empathy* Frans de Waal includes the following description of chimpanzees helping each other:

> There is in fact so much evidence for altruism in apes that I will pick just a handful of stories to drive home my point [. . .] At our primate center, we have an old female, Peony, who spends her days with other chimpanzees in a large outdoor enclosure. On bad days, when her arthritis is flaring up, she has great trouble walking and climbing. But other females help her out. For example, Peony is huffing and puffing to get up into the climbing frame in which several apes have gathered for a grooming session. An unrelated younger female moves behind her, places both hands on her ample behind, and pushes her up with quite a bit of effort, until Peony has joined the rest.
>
> (de Waal, 2009, p. 105)

De Waal presents many descriptions like this one of how chimpanzees spontaneously help each other without requiring anything in return. In *Primates and Philosophers, How Morality Evolved* he writes: 'Since expressions of sympathy emerge at an early age in virtually every member of our species, they are as natural as the first step' (de Waal, 2006, p. 28). He sees altruism as spontaneous emotional responsiveness that evolves naturally as we grow up with others.

De Waal follows here a tradition of thought that can be traced back to Smith, Darwin and Westermarck.[2] In *The Theory of Moral Sentiments*, Adam Smith writes:

> How selfish soever man may be supposed, there are evidently some principles in his nature, which interest him in the fortune of others, and render their happiness necessary to him, though he derives nothing from it except the pleasure of seeing it. Of this kind is pity or compassion, the emotion which we feel for the misery of others, when we either see it, or are made to conceive it in a very lively manner.
>
> (Smith, [1790] 1982, p. 9)

According to Smith, human beings have a natural inclination to form social bonds and relationships and to feel compassion for others. We spontaneously care for others *without* requiring anything in return.[3]

Charles Darwin also maintains that human beings, as well as many animals, naturally have a 'social instinct' and an inclination to help each other. In *The Origin of Species and The Descent of Man* ([1859, 1871] 1872) Darwin writes:

> [T]he social instincts lead an animal to take pleasure in the society of its fellows, to feel a certain amount of sympathy with them, and to perform various services for them.
>
> (Darwin, [1859, 1871] 1872, p. 472)

He continues:

> The feeling of pleasure from society is probably an extension of the parental or filial affections, since the social instinct seems to be developed by the young remaining for a long time with their parents; and this extension may be attributed in part to habit, but chiefly to natural selection.
>
> (Darwin, [1859, 1871] 1872, p. 478)

Darwin sees it as a central feature of natural selection that animals have natural 'social instincts'. Westermarck is another philosopher who is influenced by both Adam Smith and Darwin (and who also has influenced de Waal). Like Smith, Westermarck emphasises our spontaneous responsivity towards each other. And in line with Darwin he claims that the origin of our social instincts can be traced back to certain biological social patterns such as the natural relation between parent and child. In *ODMI II*, he writes: 'There is one form of the altruistic sentiment which man shares with all mammals and many other animals, namely, maternal affection' (Darwin, [1859, 1871] 1872, p. 186). He continues:

> When the young are born in a state of utter helplessness somebody must take care of them, or the species cannot survive [. . .] The maternal instinct may thus be assumed to owe its origin to the survival of the fittest, to the natural selection of useful spontaneous variations.
>
> (Darwin, [1859, 1871] 1872, p. 188)

So far I have tried to describe a certain pattern of thinking that can be traced back to Adam Smith's thoughts about sympathy, that continues with Darwin and Westermarck and that today can be seen in de Waal's conception of altruism. A central feature among these thinkers is their emphasis on the naturally social and emotional character of human life.

Altruism as consisting of a social contract

In contrast with the previously described sentimentalist conception of altruism, there is a more rationalistic conception. According to Leda Cosmides and John

Tooby, the human mind has adapted over thousands of years so that we have become the social and moral beings of today. Through evolutionary history, we have developed a mental skill of calculating how to act optimally in our dealings with others. All such dealings are formed so that they on the whole tend to enhance inclusive fitness, that is, the propagation of copies of the agent's alleles in subsequent generations.[4] Cosmides and Tooby (2008) describe this view as a social contract theory:

> Social contract theory is based on the hypothesis that the human mind was designed by evolution to reliably develop a cognitive adaptation specialized for reasoning about social exchange.
>
> (Cosmides & Tooby, 2008, pp. 69–70)

Cosmides and Tooby further describe the fitness-enhancing systems of reasoning to which our human minds have adapted as 'evolutionarily stable strategies' or 'ESS'. According to them, human social engagement has evolved because it has proved to be an evolutionarily stable strategy to cooperate with others. However, social exchanges can, according to Cosmides and Tooby, also contain uncertainty because people can cheat. A cheater is a person who does not return benefits:

> A *cheater* is an individual who fails to reciprocate: who accepts the benefit specified by a social contract without satisfying the requirement that provision of that benefit was made contingent upon.
>
> (Cosmides & Tooby, 2005, p. 1296)

The ability to detect cheaters is therefore of essential importance for the stability of the social systems of cooperation. According to Cosmides and Tooby, the human mind contains a 'cheater detection system'. The ability to reliably and systematically detect cheaters is a necessary condition for cooperation (Cosmides & Tooby, 2005, p. 1296). If we were not able to detect cheaters, that is, if we could not rely on people returning our favours, we would not be able to cooperate in an evolutionarily stable manner. In another article, Tooby et al. (2008) note that human beings tend to be more altruistic towards their family than towards other people. This can, again, be understood by considering the term 'inclusive fitness'. We care for our kin because they enhance the propagation of our alleles. There is also no risk for cheating in kinship relations because both parties in kinship relations gain from cooperating. This tendency to help our kin they define as 'kin selection'. Kin selection is, according to them, an evolutionarily stable strategy that enhances inclusive fitness. Tooby et al. also speak in this context of a 'welfare trade-off ratio' or 'WTR'. By a 'welfare trade-off ratio' they mean a mental mechanism or a 'variable' that regulates how much we ought to help others in order to gain certain benefits. This variable can be 'upregulated' when we are dealing with genetic relatives:

> [N]atural selection should have designed the human motivational architecture to embody programs determining how high one's welfare trade-off ratio

> toward other individuals should be set. [. . .] kin selection theory tells us that, all else equal, WTR should be upregulated for close genetic relatives, motivating us to help kin more and harm them less than we otherwise would.
> (Tooby et al., 2008, p. 260)

Tooby et al. (2008) further describe these systems of reasoning as occurring *nonconsciously*. They conclude that altruism is the outcome of a nonconscious mental mechanism whereby we calculate how to cooperate with others.

Tooby et al. follow a pattern of thinking that can be traced back to Thomas Hobbes' idea that human social life can be described in terms of a trade-off scenario within the frame of a social contract.[5] However, there is a notable shift in meaning from a political towards a more cognitively and biologically oriented explanation. For Hobbes the social contract was a theoretical explanation of the origins of society. When he described the state of nature as a state of war between individuals, he did not claim that this is how things have *actually* originally been. Rather he wanted to paint a *theoretical picture* in order to advance his ideas about how to achieve a peaceful society. When Hobbes talks of the *natural state* of man as a state of war where we cannot trust each other and when he describes *the social contract*, he constructs a theoretical image that is not meant to be taken as a historical or biological hypothesis. The suggestion that what is described is really a biological fact about our species and real mental functions comes later, with the evolutionary biologists William D. Hamilton (1964) and John Maynard Smith (1982). Hamilton argues that an organism's survival ought to be understood according to 'inclusive fitness'. To Hamilton, this explains why some species are altruistic. Following Hamilton's ideas on inclusive fitness, John Maynard Smith introduced the idea of 'evolutionarily stable strategies' and argued that 'kin selection' is such a basic evolutionarily stable strategy. J.M. Smith (1982) also introduced the economic principles of game theory in his reflections on evolution. Cosmides and Tooby largely follow this line of thinking.

Different levels of description

So far I have described two main traditions of ideas with regard to altruism. There is the sentimentalist perspective, beginning with Adam Smith and continuing with variations, in Darwin, Westermarck and de Waal. On the other side there is the rationalist perspective beginning with Hobbes and continuing, with alterations, in Hamilton, J.M. Smith, Cosmides and Tooby. These two traditions of ideas might appear to be sharply opposed. However, it is not evident that they are so. Rather, one can consider the two perspectives to offer two different *levels* of explanation. For instance, when de Waal states that our compassionate responses are not based on any attempt to gain benefits, Cosmides and Tooby might well agree and reply by saying that such responses of course do not *appear* to involve calculation; it is simply the case that these calculations occur *nonconsciously*. Similarly, even if the 'cheater detection system' may not be readily observable in our daily dealings with each other, Cosmides and Tooby may claim that it *underlies* all our social interactions and all our reasoning in social situations as an algorithmic process in the mind.

De Waal (2008) also discusses the fact that evolutionary descriptions can be on different levels. According to him, they can take place on two levels, namely *ultimate* and *proximate* levels of description. By an ultimate description, de Waal means an evolutionary explanation on a *general* level concerning how something is useful for the *species*. A proximate description, on the other hand, explains behaviour on an *individual* level – for instance, concerning how certain animals in a specific group behave towards each other.

This difference can also be described by the terms *biological altruism* and *psychological altruism*. By biological altruism is meant animal behaviour that enhances the survival of the species. Psychological altruism, on the other hand, refers to an animal's individual psychological inclinations that motivate the animal to help others.

One might get the impression that Cosmides and Tooby talk about biological altruism (on an ultimate level), whereas de Waal describes psychological altruism (on a proximate level). However, it is not quite clear on what level Cosmides and Tooby's description is actually placed. In order to point at this ambiguity I now discuss one of their examples.

Cosmides and Tooby refer to certain empirical observations of hunter-gatherer cultures, which they maintain will give us an idea of our unadulterated evolutionary past. With the example of how a 'primitive' !Kung San woman reasons about food exchanges, they want to point to the natural as well as universal forms of social contract reasoning:

> When Agent X provides a benefit to Agent Y, triggering the expectation in both that Y will at some point provide a benefit to X in return, a social exchange relationship has been initiated. Indeed, within hunter-gatherer bands, many or most reciprocity interactions are implicit.
>
> (Cosmides & Tooby, 2008, p. 72)

To illustrate their point, Cosmides and Tooby then offer the following quotation from Nisa, a !Kung San gatherer from Botswana who was interviewed by Marjorie Shostak[6]:

> If a person doesn't give something to me, I won't give anything to that person. If I'm sitting eating, and someone like that comes by, I say, 'Uhn, uhn. I'm not going to give any of this to you. When you have food, the things you do with it make me unhappy. If you even once in a while gave me something nice, I would surely give some of this to you'.
>
> (Shostak, 1981, p. 89)[7]

Commenting on this passage, Cosmides and Tooby then add:

> Nisa's words express her expectations about social exchange, which form an implicit social contract: If you are to get food in the future from me, then

> you must share food with me. Whether we are San foragers or city dwellers, we all realize that the act of accepting a benefit from someone triggers an obligation to behave in a way that somehow benefits the provider, now or in the future.
>
> (Cosmides & Tooby, 2008, p. 72)

This example of the !Kung woman's reasoning can seem to show that there is a basic, underlying pattern of social contract reasoning in all our social encounters.

However, Shostak's study of the !Kung has also received some criticism. Gupta and Ferguson (1992) argue that Shostak's study is problematic in several ways. One problem is that the image of a whole culture is based on a single person's words: 'The individual, Nisa, is granted a degree of singularity, but she is used principally as the token of a type: "the !Kung" '(Gupta & Ferguson, 1992, p. 15). Gupta and Ferguson also question Shostak's suggestion that the !Kung represent an untouched, primordial people. They refer to an article by Mary Louise Pratt (1986) where she notes that the isolated life of the !Kung people is the result of centuries of colonialist violence. Gupta and Ferguson quote Pratt:

> What picture of the !Kung would one draw if instead of defining them as survivors of the stone age and a delicate and complex adaptation to the Kalahari desert, one looked at them as survivors of a capitalist expansion, and a delicate and complex adaptation to three centuries of violence and intimidation?
>
> (Pratt, 1986, p. 49)

According to Cosmides and Tooby, anything that is done among hunter-gatherers exemplifies the basic natural features of all human reasoning. Pratt's criticism of Shostak's study suggests, however, that the portrayal of 'primitive' tribes can be misleading partly because it is taken for granted that such tribes have no history, that they have always lived like this. Gupta and Ferguson in their turn argue that the image of the 'primitive' tribe is problematic because its basis is too narrow (only one person in Shostak's case).

Not only is it problematic to take a single person's words to illustrate a whole culture's attitude. My impression is also that Nisa's words can be understood in several ways. It is not self-evident that she is neutrally describing a general pattern of exchanges of benefits. To me it seems that the !Kung woman is embittered, as anyone can sometimes be. She seems to be fed up with sharing her food with others who apparently have taken advantage of her generosity. The fact that people sometimes grow bitter and cynical does not prove that this is how we all nonconsciously 'function' socially. The mere fact that Nisa lives a life as a hunter-gatherer does not imply that her thinking is more natural and immediate than ours. Nor is the fact that all people around the world *sometimes* become bitter proof of the absolute basically transactional function of social life.

What I have suggested here is that Cosmides and Tooby's rationalistic perspective does not actually describe an underlying cognitive mechanism. Rather their perspective is expressive of certain kinds of reductive descriptions of human life.

These reductions take several forms. First, there is a tendency to take examples from ordinary human life situations and then claim that these examples point towards a general cognitive mechanism. By taking an example of how a woman from a 'primitive' hunter-gatherer culture reasons, they create the impression that they describe a basic pattern in human reasoning. There is also a shift in meaning by the use of a reductive language. Cosmides and Tooby tend to use a strategic vocabulary when describing human relationships. They talk about 'kin selection', 'cheater detection' and 'welfare trade-off ratio' when describing our care for family members. These expressions are expected to describe general underlying behavioural patterns. Our ordinary relational words such as 'child' or 'parent' are supposed to gain their meaning from these underlying strategic concepts. These expressions further the impression, as I mentioned earlier, that human relationships are based on a strategic calculus.

I now return to discuss the sentimentalist perspective on altruism. I will suggest that also this perspective builds on certain problematic conceptions of human understanding.

Sympathy, empathy and retributive emotions as consisting of analogical imagination

One idea that Adam Smith, Darwin, Westermarck and de Waal have in common is a conception of emotions as reciprocal or (using Westermarck's term) 'retributive'. This is again connected with a certain conception of the nature of interpersonal understanding. As I already mentioned, all four authors would emphasise our spontaneous inclination to care for others. However, all four also maintain that this inclination can be explained as deriving from a certain function embedded in the mind, even if it is not a computational function.

According to Adam Smith, our care for other people is connected with our capacity to imagine ourselves in the other's situation. This can be seen in his ideas on sympathy. By 'sympathy', Smith does not mean simply compassion, but any kind of imaginative experience of another person's feelings that cause us to be emotionally moved by the other. This is, in modern terms, called 'empathy'. The explanation for our spontaneous reactions of compassion must be sought in the functioning of our minds and in our capacity to imagine how others feel:

> As we have no immediate experience of what other men feel, we can form no idea of the manner in which they are affected, but by conceiving what we ourselves should feel in the like situation. Though our brother is upon the rack, as long as we ourselves are at our ease, our senses will never inform us of what he suffers. They never did, and never can, carry us beyond our own person, and it is by the imagination only that we can form any conception of what are his sensations. Neither can that faculty help us to this any other way, than by representing to us what would be our own, if we were in his case. It is the impressions of our own senses only, not those of his, which our imaginations copy. By the imagination we place ourselves in his situation, we

> conceive ourselves enduring all the same torments, we enter as it were into his body, and become in some measure the same person with him, and thence form some idea of his sensations, and even feel something which, though weaker in degree, is not altogether unlike them.
>
> (Smith, [1790] 1982, p. 9)

Smith's explanation of moral sentiments consists in an argument from analogy. Sympathy is the effect of our capacity to imagine ourselves in the other's shoes. This effect can in turn make us feel compassion for the other person's misery. Westermarck argues along similar lines as Smith. According to him, morality is the outcome of 'retributive emotions':

> Our retributive emotions are, of course, always reactions against pain or pleasure felt by ourselves; this holds good of the moral emotions as well as of anger, revenge, and gratitude. [. . .] That a certain act causes pleasure or pain to the bystander may be due to the close association that exists between these feelings and their outward expressions. The sight of a happy face tends to produce some degree of pleasure in him who sees it; the sight of the bodily signs of suffering tends to produce a feeling of pain. In either case the feeling of the spectator is due to the fact that the perception of the physical manifestations of the feeling produces the feeling itself on account of the established association between them.
>
> (ER, p. 96)

According to Westermarck, when I see another person's bodily expressions, this causes me to associate these expressions with certain feelings, and thereby I experience these feelings myself. This analogical mechanism of imagination explains, according to him, why we are inclined to help another person when he or she is in pain. However, Westermarck also suggests that this spontaneous emotional responsiveness to others develops through habit. In this sense he does not merely consider compassion to derive from a mental capacity of analogical imagination. Rather he claims that our inclination to imagine how other people feel grows out of our habits of shared relationships. It is, according to him, important that we get used to sharing our life with each other, for how our emotional responsiveness also develops.

De Waal's ideas are largely along similar lines as those of Adam Smith, Darwin and Westermarck. As already suggested, according to de Waal our inclination to care for each other is grounded in the fact that we are born into social relationships. But he also thinks our social responsiveness depends on a mental mechanism of imagination. According to him, both humans and apes have a natural capacity to imagine how others feel, a capacity that is enhanced by our habitual ways of living together.

De Waal, who is more biologically oriented than Adam Smith and Westermarck, also describes empathy as at first originating from a bodily reaction that eventually develops into a full-fledged capacity to imagine another's perspective.

According to de Waal, then, first we simply have a tendency for what he calls 'emotional linkage' or 'emotional contagion'. Eventually we also develop the capacity to see that others have different perspectives on reality. This is, according to him, part of what it means to feel empathy:

> When the emotional state of one individual induces a matching or closely related state in another, we speak of 'emotional contagion' [. . .] With increasing differentiation between self and other, and an increasing appreciation of the precise circumstances underlying the emotional states of others, emotional contagion develops into empathy. Empathy encompasses – and could not possibly have arisen without – emotional contagion, but it goes beyond it in that it places filters between the other's and one's own state.
>
> (de Waal, 2006, p. 26)

According to de Waal, this capacity for empathy is *not* something that can be explained as deriving from social competition, but it derives from a need for *cooperation*. Human beings as well as apes have a natural inclination to become emotionally affected by others because we have a natural need to cooperate. At the same time, however, our emotional responsiveness enhances our natural inclinations to cooperate. One can consider the relation between empathy and cooperation as a continuous circle:

> I am personally convinced that apes take one another's perspective, and that the evolutionary origin of this ability is not to be sought in social competition, even if it is readily applied in this domain [. . .], but in the need for cooperation. At the core of perspective-taking is emotional linkage between individuals – widespread in social mammals – upon which evolution (or development) builds ever more complex manifestations, including appraisal of another's knowledge and intentions.
>
> (de Waal, 2006, p. 72)

This 'emotional linkage' eventually leads to a more advanced capacity to imagine another's perspective, which de Waal calls empathy. Even if I think there is much truth in de Waal's ideas on how our responsiveness to each other – and our care for each other – grow out of our habits of sharing each other's life, I also think it is problematic to claim that, at the bottom of our responsiveness, there is some such thing as an analogical mechanism of imagination. De Waal describes the empathic mechanism as follows:

> [A]t the core of the empathic capacity is a relatively simple mechanism that provides an observer (the 'subject') with access to the emotional state of another (the 'object') through the subject's own neural and bodily representations. When the subject attends to the object's state, the subject's neural representations of similar states are automatically activated. The closer and more similar subject and object are, the easier it will be for the subject's

> perception to activate motor and autonomic responses that match the object's (e.g., changes in heart rate, skin conductance, facial expression, body posture). This activation allows the subject to 'get under the skin' of the object, sharing its feelings and needs, which embodiment in turn fosters sympathy, compassion, and helping.
>
> (de Waal, 2006, pp. 37–8)

I agree with de Waal (and Smith and Westermarck) that it is a common feature that we are emotionally affected by others and that a kind of emotional contagion often occurs. But the question is whether this can be understood as expressive of an analogical mental mechanism, as they describe it. And the question is also whether compassion can be understood in this sense as deriving from analogical imagination.

The assumption that compassion is based on a cognitive capacity for analogical imagination reflects the assumption that I understand the other person's emotions by *first* understanding what the emotion means in my own case, and by then projecting these feelings onto the other person. Such an idea rests on a classical form of philosophical scepticism where the idea is that we can only really know what we feel ourselves. In other words, the idea that compassion is based on the use of an analogical method of imagination (i.e., empathy) is expressive of a tendency to assume that the second-person perspective (i.e., how we respond to other people) is dependent on the first-person perspective, when we talk about sensations. This reflects the assumption that human understanding takes place in the individual being's mind, consisting of an inner *image* or *impression*. Such an assumption rests on a conceptual confusion regarding the concept of understanding, turning the concept of understanding into something *private* in the mind. However, the concept of understanding, including what we mean by such words as compassion, care and considerateness, has meaning centrally in our interactions with each other. We ask others for help, we criticise someone for being careless or callous, we admire someone for being so considerate, etc. That is, when we say that a person is considerate or acting in a compassionate way, we do not refer to a certain inner state or feeling but to how the person acts towards others.

We do sometimes imagine how we would feel in another person's situation, and these thoughts can sometimes be connected with compassion; but this is merely one form our care for another person can take. Our care for another person can equally well take the form that we do *not* use our imagination. To take an example, you see a child hit by a car and you run instantly to help the child. The only thing that passes your mind might be something like 'Oh my God!' That is, often we simply react to suffering and no imagination is needed. But sometimes we might also imagine things. For instance, in the evening after the accident you cannot stop thinking about the child. Your thoughts about the child can also be entwined with how you think about your own children. Here one could say that a kind of analogical imagination occurs. However, this analogical imagination is in itself already a kind of moral response; it is one way you are shocked and can't stop worrying. The fact that reactions of this kind can occur does not show that this is a basic cognitive mechanism that enables us to understand other people. We can respond with compassion in many different ways to other people's suffering, but this does not mean that *one* of these ways of responding is the basis for all other kinds of responses. It can also be noted that, in the case earlier, engaging in analogical imagination is a reaction *after* the accident. It is, in this case, not something that precedes, and motivates, the helping action.

My intention here has not been to claim that it would *never* be of any importance for our understanding that we can react in an analogical way to other people's emotions. What I do want to argue is that it is problematic to claim that our care for others builds on some sort of unitary cognitive function. Both animals and human beings can respond with care in many different ways to other people's suffering, but this does not mean that *one* of these ways of responding is the basis for all other kinds of responses. Furthermore, it is also important to acknowledge that there are human ways of caring that cannot be understood or described from a general perspective on animal behaviour. The problem with Adam Smith, Darwin, Westermarck and de Waal is their conception of how analogical imagination works and of its importance. They assume that it is a general and basic cognitive method or function that enables us to understand other people.

Natural reactions, language and a shared life history

So far I have contrasted two theoretical perspectives on altruism; one rationalistic (transactional) and another sentimentalist (and analogical) in kind. These perspectives can appear to be quite opposite. However, I have suggested that one can consider them as being situated on two different levels of explanation. Still, both perspectives also reflect certain problematic conceptions of interpersonal understanding. In what follows I will argue that both perspectives suggest that human social life can essentially be described in instrumental terms, even though they have somewhat different conceptions of this instrumental basis of human social life.

Consider the following description by Flack and de Waal (2000a) of food-sharing among chimpanzees:

> Food sharing is known in chimpanzees [. . .] It is an alternative method to social dominance and direct competition by which adult members of a social group distribute resources among themselves. Most food sharing requires fine-tuned communication about intentions and desires in order to facilitate inter-individual food transfers. [. . .] [T]he reciprocity hypothesis – proposes that food sharing is part of a system of mutual obligations that can involve material exchange, the exchange of social favours such as grooming and agonistic support, or some combination of the two.
>
> (Flack & de Waal, 2000a, pp. 4–5)

Flack and de Waal suggest that food sharing is an example of what they call 'the reciprocity hypothesis', that is., a system of exchanges of favours. This description resembles Cosmides and Tooby's example of the !Kung San woman discussed earlier. The main difference is that Cosmides and Tooby's example concerns a *negative* situation where food exchange does not work well, whereas Flack and de Waal describe a *positive* situation of smoothly working food exchange.

Though I do think there is much truth in the claim that the human species is a cooperative species, there are problems with the way our care for each other is described as an exchange of favours. Such a description does not show how our practical, and nonpractical, ways of living have meaning as an *integral* part of our close relationships. Neither does it show how our care for each other takes a *human* form. Consider, again, what it means to eat together. Human beings do not simply exchange food; we eat *together*. That we eat together often displays the self-evident way in which our lives are intimately woven together in multifarious ways. At the same time, the fact that we share meals is expressive of our sense of responsibility for the other. This is a responsibility that cannot be understood without acknowledging the importance of our shared presence for each other in eating together. This shows, for instance, in that we often spontaneously *talk* with each other and *listen* to each other while we eat.

According to Flack and de Waal, communication is used in food sharing in order to 'facilitate inter-individual food transfers'. I agree that talking while eating can be of practical importance. However, the fact that we talk at meals also often has a nonpractical character. We simply talk because we are together. We are often spontaneously drawn into conversations while we eat with others. This may be something none of us *decides* to do; and it need not have any further purpose. This spontaneous presence is in itself expressive of moral sensitivity towards the other – my ability to be drawn into the other person's life, to become interested in his life and to take his life seriously.

Of course, there can be situations where people do not talk with each other while eating. In some families it might be even more of a rule that they do not talk while eating. One reason can be that the married couple is deeply fed up with each other. In such situations, a cold manner of talking merely in order to 'facilitate inter-individual food transfers' might also take place. However, such ways of eating are not any more basic or more natural than are our ways of enjoying eating together. Nor do such ways of eating and talking *explain* the multifarious ways we share a life with each other and find meaning in being together.

Ludwig Wittgenstein writes: 'Commanding, questioning, storytelling, chatting, are as much a part of our natural history as walking, eating, drinking, playing' (Wittgenstein, [1953] 2009, §25). Wittgenstein suggests here that it is problematic to think of human nature as if it had an 'essentially natural' part consisting of practical activities essential for survival, whereas our nonpractical activities would not be as essentially natural but 'cultural' or 'psychological' features that enhance our practical cooperation. Another thing he suggests is that it is problematic to think of language as basically a practical tool. Wittgenstein mentions such things as chatting, storytelling and playing – that is, ways of talking and doing things that do not have a specific goal but are rather forms of being with others. The meaning of our practical activities can often not be separated from the way these are entwined in activities and ways of talking that do not have a practical character. A further thing Wittgenstein's remark suggests is that a large part of our activities have meaning because these activities are integral to the fact that we stand in personal relationships. The practical help and care that we give each other is an integral part of a larger pattern whereby we acknowledge each other *as persons*. In this sense our physical needs largely get their form and meaning from the fact that we live a shared life with others, and a shared life full of talk. Thus, what it means to understand a child's physical needs cannot be separated from our *acknowledging* this child as someone to be with also in other ways, someone to eat together with, someone to chat with, quarrel with, and tell stories to, someone to help and teach, someone with whom we share a future life and a history.

Conclusion

From my previous discussion one might get the impression that I claim that human social life cannot be described from an evolutionary perspective at all. This has not been my intention. I agree with evolutionary theorists that many social patterns of human life enhance the survival of the species. I do also think there are interesting similarities between the social behaviour of many animal species and that of human beings.

However, I have argued that there are certain problems in both the rationalist and sentimentalist perspectives on altruism. Both perspectives build on an instrumental conception of human social life. This shapes how they describe examples of interaction. It also shapes their choice of words when describing interpersonal relationships. Further, I have claimed that the instrumental conception of human social life is connected with two kinds of conceptual confusions regarding interpersonal understanding. According to Cosmides and Tooby, human understanding is a computational function in the mind. However, it is an essential feature of what we mean by human understanding that we can comment on – and criticise – each other, that we can teach others and show others how to do things, that we can condemn others or admire others for their actions, etc. I mentioned earlier that Cosmides and Tooby talk about 'cheater detection modules' and 'welfare trade-off ratio'. These are not only highly specialised concepts; they also reflect both the instrumental conception of human social relationships and the assumption that human understanding consists in an algorithmic function; a function that 'works' in us regardless of what we think or say about it. They suggest that these concepts refer to a deeper, natural, level of our relationships. I have questioned this assumption.

Darwin, Westermarck and de Waal, on the other hand, do acknowledge that animals are born into close relationships. In this sense they acknowledge that close relationships form and give meaning to our practical actions. However, even if their descriptions are often detailed and insightful, they still tend to describe interpersonal relationships as if they basically had an instrumental purpose. This is partly reflected in their comparisons between animal and human behaviour. This is one reason for Westermarck's and de Waal's use of the concept 'altruism', or, in Darwin's case 'social instincts'. Partly their instrumental conception of interpersonal relationships is also reflected in their choice of vocabulary when describing certain behaviour, such as when de Waal talks about 'food exchange'. Further, I have also claimed that the sentimentalists assume that interpersonal understanding works according to a mental mechanism of analogical imagination. This is reflected in their use of concepts such as 'retributive emotions' and 'empathy'. I have suggested that this conception of interpersonal understanding also is problematic.

Does my criticism imply that one can *never* give an evolutionary explanation of human social life? Is it *never* all right to compare human beings and animal behaviour? This has not been my intention to claim. Clearly it can be important to study animal social life and to sometimes see connections with human social life. However, such a description can also have its limitations, which is reflected in the predilection of certain concepts. A too general use of concepts like 'cooperation' or 'transaction' easily leaves out important aspects of what it means for human beings to share a life and thus also what it means to care for another. Further, I have argued that it is problematic to assume that our emotional responsiveness to each other has evolved as a useful cognitive tool that enhances cooperation. In order to see how our life has a natural form, one must acknowledge the *human* character that our life takes. This is reflected in our use of words. The words with which we, in everyday life, refer to 'biological' relations and functions, such as eating, sex, fights, begetting children and caring for them, are imbued with meanings that go beyond both biology and strategic transaction.

Notes

1 I have discussed these questions previously in Gustafsson (2012) and (2014).

2 The relation between Smith, Darwin and Westermarck is also discussed in Chapter 10 by Otto Pipatti.

3 Smith does not specifically use the word 'altruism', as no such word was in use at the time of his writing. The French term '*altruisme*' was first introduced into philosophical theory by Auguste Comte in his *Catéchisme Positiviste* (1852). This term was translated as 'altruism' in Richard Congreve's English rendering of the text *The Catechism of Positive Religion* (1858). Smith's ideas on compassion are, however, similar to Comte's ideas on altruism, inasmuch as Comte does not think of altruism as being based on self-interest.

4 In *The Adapted Mind* (1992) Cosmides and Tooby distinguish between two senses of 'fitness'. They write:

> Previously, evolutionary biologists spoke of a design's "Darwinian fitness": its effect on the number of offspring produced by an individual who has the design. But since Hamilton, one speaks of a design's "inclusive fitness": its effect on the number of offspring produced by an individual who has the design *plus* its effects on the number of offspring produced by others who may have the same design – that individual's relatives – with each effect discounted by the appropriate measures of relatedness.
>
> (Cosmides and Tooby 1992, p. 168)

When Cosmides and Tooby talk of fitness they mean 'inclusive fitness'.

5 See, for instance, Hobbes ([1651] 1985), *Leviathan*, chapters 13–15.

6 The work by Shostak that Cosmides and Tooby refer to is *Nisa: The Life and Words of a !Kung Woman* (1981).

7 Shostak is quoted in Cosmides and Tooby (2008), p. 72.

15 Westermarck, sympathy and natural selection

Hannes Nykänen

Introduction

The concept of sympathy has become important in current discussions of evolutionary psychology and moral philosophy. Some commentators think that sympathy is a prerequisite for morality, whereas still others think understanding the role of sympathy in primate behaviour will cast light on the evolution of morality (Darwin, 2010; de Waal, 1996; Flack & de Waal, 2000a; Sober & Wilson, 2003; Thompson, 2001; Westermarck, ER). Others are sceptical about any reference to moral psychological explanations in biology (Nesse, 2000; Pinker, 2012; Zahavi, 2000), and still others think that the only instance of unselfish behaviour is to be found in collective togetherness (Haidt, 2012). Then there are, of course, those moral philosophers who deny that moral acting could be motivated by emotions and those who think that feeling is essential to moral acting, but that this feeling must be a moral feeling (Bagnoli, 2011; Morrisson, 2008).

The current discussion on sympathy encompasses many big topics: methodological issues, the nature of morality, the classical dichotomy between reason and emotion, egoism and altruism, individuality and collectivity and the role of group selection, to mention the perhaps most important ones. Despite these differences of opinion, these discussions have promoted a certain view of not only sympathy but also of what morality is about. Within the dominating, naturalist camp, the argument concerns the evolutionary role of sympathy and its relation to selfishness and altruism, with one major issue being whether group selection, and so altruistic motivations, can play any important role in natural selection. Philosophers belonging to the more philosophically orientated 'transcendentalist' camp reject naturalism by reference to what they take to be the cognitive nature of morality; to forms of moral reasoning that in their opinion cannot be identified with biological functions.

Nevertheless, naturalists and transcendentalists share certain presuppositions about emotion and reasoning. The goal of the present paper is to reveal the inadequacy of these shared presuppositions by looking at emotions and moral reasoning from a new angle – an angle that will reveal the common roots of naturalism and transcendentalism. My intention is to show in what sense sympathy is a fundamental form of moral understanding not reducible to subjective feeling and why natural selection cannot put pressure on it.[1]

Much has been said about the way the naturalistic assumptions in evolutionary psychology and neuroscience are too simplistic and narrow (Descombes, 2001; Noë, 2010) or constitute a mistaken way of using language and reasoning (Bennett & Hacker, 2003). These forms of criticism, although certainly to the point in many ways, do not give us a clear enough idea of what the problems with naturalistic accounts of moral understanding are. My aim is to show why it does not make sense to think that natural selection affects moral behaviour. This involves showing how sympathy differs from all other feelings in that it is not self-related but in an unspecific way other-oriented.[2] Moreover, I will show that this has a much more radical meaning than is usually thought.

According to evolutionary psychologists, the key issue is to explain how biologically determined psychological reactions – for it is among these that moral 'behaviour' is supposed to belong – have been modified and diversified by culture and assess whether or not group selection has a significant evolutionary effect. Evolutionary psychologists generally think that morality became possible only because of the growth of intelligence. Intelligence is seen as the source of freedom and thus as the condition of the possibility of morality (see, for instance, Dennett, 2004, p. 260). This idea would be very close to Kant's basic idea if it were not for the fact that for Kant it was *practical* reason that was capable of freedom and morality – not, as is the case with evolutionist thinkers, *theoretical* reason. In Kant's view, theoretical reason cannot even recognise the question of freedom just because it is algorithmic, that is, it traces step by step the 'mechanism of nature'. (Many remarks in Kant's work point in this direction, but see especially Kant, 1979, pp. 15–18, 1996, pp. 43–6, 129; or AA3:16–19 and AA5:28–30.)

I will not discuss the merits and problems of different views of sympathy and empathy (see Blum, 2011; Flack & de Waal, 2000a; Haidt, 2012; Prinz, 2011; Thompson, 2001; van der Weele, 2011), but instead focus on a mostly neglected or misunderstood aspect of it – namely the way sympathy falls completely outside the opposition between egoism and altruism and, instead, involves a desire for the other as such. This makes the idea that morality could be targeted by natural selection impossible to sustain.

The focus of my critical discussion is directed at Edward Westermarck's sentimentalist theory of morality. However, I will begin with a critique of some central aspects of the main target of Westermarck's criticism, namely Kant's ethics. Both Westermarck and contemporary evolutionary psychologists have in fact taken over some of the most problematic elements of Kant's account of ethics. After the section on Kant, I will outline my own view of sympathy and then take issue with Westermarck's sentimentalist account of ethics.

Kant, feeling and nature

The core of Kant's conception of feeling is that it does not contain any understanding. The fact that we can use our reason as a tool for satisfying the urges connected to feeling should not be confused with this. Feeling is one among different kinds of signals from the bodily mechanism. An itch is an itch, thirst is thirst and fear is fear; there is nothing in these sensations and feelings themselves that *can* be understood in a way that goes beyond simply feeling them. They are defined as *lacking* understanding and are in this sense 'blind and slavish' (Kant, 1996, p. 143; 2000, p. 12; or AA6:212, AA5:118). Feelings contain no intelligibility because they are only urges to get an object; they have no content apart from urging us to capture the object we urge for. One could also say that Kant, adopting the standard philosophical stance, is 'forced' to separate the bare urge from the meaning, subjectivity from objectivity. Whatever thoughts go with, say, fear, they 'must' be simply objective and rational and hence distinct from the feeling of fear.

Feeling an urge for something and satisfying this urge are obviously completely self-regarding, and if one uses rationality to satisfy the urge, this does not, of course, make the action any less selfish. In fact, it has often been held that selfishness is criterional for the rationality of an action. Although Kant, of course, tried to fend off egoism with his moral theory, it is important to note that his conception of feelings, inclinations and rationality create a fundamentally egoistic framework. The importance of 'altruism' arises as a response to the unacknowledged, generalised egoism. This is backward because egoism can arise only *within* moral understanding. My intention in this chapter is to show that morality is constituted by something that is morally prior to the dichotomy between egoism and altruism, namely 'sympathy'.

Even though sympathy cannot in Kant's view constitute a moral motive, it still has a decisive role in his ethics. I will try to locate some of the confusions and insights in Kant's project and show how they hint towards the crucial moral role of sympathy.

Kant's insight that moral problems (fundamentally: the question of freedom[3]) do not concern theoretical reason shows itself in two ways: partly in that he takes practical reason to be crucially different from theoretical reason and partly in that he gives feeling a decisive role in ethics (see Kant, 1996, pp. 17, 142–5, 2000, p. 160 or AA5:7, 116–19, and AA6:399–400). I will here focus on the latter aspect, although the two aspects are closely related.

The sense in which moral feeling differs from bodily feelings is that it is not prompted by natural, bodily inclinations but by moral understanding – more specifically by reverence for the moral law. Yet, because it is a feeling, moral feeling is nevertheless bodily. This is an important qualification of ethics as something that concerns us as natural, feeling beings. Moral feeling or respect involves the humiliation of self-conceit caused by an a priori insight about the authority of the moral law. Respect is caused by practical reasoning, and it humiliates selfish impulses (1996, pp. 93–4 or AA5:73).

Kant was right to worry about allowing self-regarding impulses to determine morality, but he did not see that a feeling's being 'unpleasant' does not make it any less egocentric – a theme that Freud was to make much of. In fact, there would be a lot to say about the way self-destructive and self-punitive impulses are connected to ruthless acting and mental disorders. Humiliation is as self-regarding as any other feeling. The 'positive' aspect of moral feeling is equally egocentric, for it is a 'self-approbation' in recognising in *oneself* the ability to be moved by the moral law alone (1996, pp. 101–2 or AA5:80–1). Moreover, Kant's idea about the free, rational subject that legislates for *itself* in awe of the moral law is egocentric too. Despite his efforts to avoid it, Kant's view falls prey to the egocentrism that is typical for both moral philosophy and evolutionary psychology. (Dennett, 2004, pp. 193 ff., thinks that egoism is, if 'moderate', a moral virtue, whereas Taylor (2011) fails to see the egoism of the outlook she offers.[4])

Kant's effort to incorporate feelings into his account of morality creates immense problems: (1) In the coming together of practical reason and feeling something noumenal, that is, morality, makes contact with something phenomenal, that is, feeling. How is that possible? (2) How should one distinguish between this very special moral feeling and other feelings that are 'purely' natural, for both kinds of feeling include bodily sensations? (3) Why does the moral law, being completely formal, inspire such a strong feeling – perhaps the strongest of all?

Kant's view implies, and he sometimes says, that feelings are unrelated to morality and acquire moral relevance only if they coincidentally interfere with the moral law. In line with this, duty involves autonomy, the ability to rule over one's inclinations (2000, pp. 146, 166 or AA6:380–1). But Kant also speaks as if feelings could 'rebel' against the moral law, and he even calls them the 'evil principle in a human being' (2000, pp. 148, 191 or AA6:383, 440). Both assumptions seem incoherent. If feelings are simply bodily, how can they *violate* the moral law? Would they not in that case be more like a natural evil, like an earthquake, rather than evil in the moral sense? If again feelings are taken to be in themselves evil, then they clearly cannot be simply natural. Clearly, too, neither practical nor theoretical reason can be the source of moral temptation. However, Kant says that the 'mind' forms evil maxims (2000, p. 166 or AA6:408) and that it is the 'sensible being endowed with reason' that is accused, by conscience, of immoral acting (2000, p. 189 or AA6:439). But because evil cannot be attributed to either feelings or reason the way he accounts for them, his view cannot consistently account for the possibilities of moral temptation and evil.

The main problem in Kant's account is, first, that he thinks feeling is devoid of any understanding. This in fact also goes for moral feeling; it 'yields no cognition' (2000, p. 160 or AA6:400). The second problem is that he thinks that feeling is necessarily self-regarding. Both contemporary evolutionary psychologists and, in the end, Westermarck endorse these claims. The previously mentioned outlined difficulties in Kant's account are of a kind that cannot be amended by interpretation. For instance, Morrisson (2008) gives a very plausible interpretation of the way Kant thinks that moral feeling is an executive and hence subjective – not legislative and objective – power of moral acting. Because we are sensible natures, we also need feeling as an incentive to moral acting. Moral feeling is unique in that it is not a response to the representation of pleasurable objects, but to the intellectual insight of the authority of the moral law (Morrisson, 2008, pp. 135 ff.). However, Morrisson's interpretation does not address the fundamental problems that Kant's account faces (nor does Morrisson claim that it does).

Kant's constant wavering between saying that feelings are pure nature and hence not evil as such and saying that they are evil is probably influenced by a deeply entrenched, often religiously underpinned, tendency in our culture to view nature as something dark and evil. (Consider what Freud and psychoanalysts have said about what they call the *id*. See also Williams, 1996, p. 255.) This same tendency probably also accounts for the fact that Kant's claim that feelings cannot have positive moral relevance has been so largely accepted.

Sympathy

If we think of feelings as messages of bodily urges, it makes no sense to think that they would contain any notion of caring for the *other*. However, and as I will show, sympathy involves an engaged appreciation of the other *as such*. This is not an idealised description, but simply what 'sympathy' *means* – what it means to have feelings for the other. Imagine a case where one person abuses another person. It is conceivable that even after having abused the victim in any imaginable way, the abuser could still long for the company of the abused. This shows in a backward kind of way that *self-regarding urges and satisfying them leaves the longing for the other untouched.*

Without going further into the moral issue of abuse, it is important to notice that corrupt forms of longing for the other can exist only if there is such a thing as longing for the other just because she is other. If we had no conception of caring for each other apart from the way we can satisfy our own and each other's self-regarding needs, there could be no corrupt forms of longing for the other, no conception of abuse. In corrupt forms of longing for the other, one desires the other for her own sake and then corrupts this desire by trying to *make* her (by way of force, threat, pressuring or seduction) to be in one's company. (A familiar case of this is the legendary trope of the villain keeping the beautiful woman hostage out of 'love'.)

These observations point at the sense in which sympathy is not a desiring in the egocentric sense of the word. 'Desiring something' in this sense involves that the desirability of the object depends *only on my desire*. But sympathy is not only about my desire, for the other's desire for me is part of my desire. Still, it is not pertinent to speak about mutual desire here.

That the other matters *qua* herself means that she has an importance that is *unrelated* to self-regarding urges. The problem of altruism and egoism arises as a response to our difficulties with sympathy. Just to indicate what this means, a common form of egoism and corruption of sympathy involves trying to buy another person's sympathy, ignoring the fact that this is impossible.

Someone might still wonder why longing for the company of another person could not be just one of the many pleasurable feelings that the brain creates in order to further the organism's well-being and fitness. Here it is important not to confuse things. Sympathy has bodily aspects to it, but this does not mean that it can be *understood* as a concern for purely self-regarding urges. Hunger and fear could, with some caution, be used as examples of such urges.

Self-regarding feelings cannot go against my interests, for they are expressions of my interests. They can, however, malfunction so that a certain urge turns into something bad for me. It is also possible for two urges to run into conflict and for one of them to silence the other. In such cases it is still quite clear that the urges, feelings and dispositions in question are wholly self-regarding.

By contrast, sympathy concerns my whole being; it does not inform me about bodily interests, but it can, as we all know, ignore them to a surprising degree. If I lacked sympathy, I would not for that reason be lacking any information relevant for preserving my body, whereas if I lacked the feelings of thirst, tiredness, caution, fear, etc., I would perish. To say that a feeling or disposition is body-preserving and so self-regarding means among other things that it is a matter for the creature that has them to manage them. By contrast, sympathy is the feeling that *feels* the other in a way which is *unconnected* to the urges that promote the preservation of one's body.

If one were to place the concept of sympathy in an evolutionary perspective, one might say that we are talking about sympathy if it makes sense to think that a feeling implies that an organism longs for the other and desires to be with her. Conversely, if a feeling for the other has the character of a *need*, it is not a feeling of sympathy, simply because it is self-regarding. Sympathy is what we here call the aspect of longing for the other as such, whereas needs, etc., are something else. Although we may take this longing to heart to different degrees, this does not mean that egocentric processes could gradually turn into sympathy (see Dennett, 2004, pp. 51, 217, who entirely misses this point).

Understanding one's feeling of sympathy as a desire for the other is inseparable from wanting to understand the other. The aspect of understanding is thus characteristic of sympathy. Sympathy is not a blind craving for the other's presence. It may be hard to say whether fish that live in shoals have sympathy or whether they simply need each other for survival, but it is clear that dogs, chimps and dolphins have sympathy for each other. To be prepared to say this *means* that one is prepared to view the interactions of these animals in the light of the concept of understanding. In human terms, the 'depth' of the shared understanding that is opened up in sympathy depends on how much one dares to be open with the other, how much one dares to share feeling and understanding. Thus, sympathy is not something that one can have as a private feeling; it is more like two persons looking at each other. This cannot be accounted for as two persons having private visual impressions of each other.[5] Moreover, sympathy is not merely an established fact but contains in itself a desire to learn to know the other more and more. Sympathy is also a task and a challenge.

Aggression and hostility do not involve lack of sympathy but overrule sympathy. If I were to face a murderer, looking into her eyes would not be like looking into the sensory devices of a robot. I would look into the eyes of a human being who would have to overrule her sympathy in order to kill me. What I would see is someone who is going to be haunted for the rest of her life for killing me. (Of course, I would hardly be thinking of such things, but I would be standing in a relationship with someone who is going to go through that.) This is the sense in which conscience presents an inescapable, not an optional, perspective on the other. There are, of course, many cases where conscience *seems* to be absent but, without entering that discussion here, none of them cause problems with what I have said – quite the contrary.

On my terms there would have to be at least some sense with speaking about 'daring to be open' in connection to animals, and unsurprisingly it does. The animal world is full of events where animals are very reluctant to approach each other for different reasons and yet it is apparent that they have a desire to approach each other, even taking risks in doing that.[6]

My description of sympathy differs from standard descriptions in that I do not see sympathy as a subjective feeling directed at an object. Moreover, in my view sympathy is not merely feeling but also understanding – but not in the sense of computational or theoretical thinking. Thus, sympathy cannot be conceived of as a combination of the standard concepts of feeling and cognition. How then should one think of it? One way of emphasising what I take to be the crucial feature of sympathy is to say that in sympathy two persons (or more), as it were, *are within sympathy*. Sympathy cannot be made sense of as a mutual feeling if this means that initially my sympathy for you is isolated from your sympathy for me.

If sympathy was mutual, that is, a transaction, it would have to be completely absent before the assumed transaction. Before the transaction I would be completely neutral with respect to sympathy, for at this stage the whole issue whether I am going to confer or withhold sympathy has not yet arisen. In this kind of situation the only way of making my sympathy known to the other would be by somehow signalling it. If I withheld sympathy, this would have to be shown by some kind of negative sign that differs from the assumed, 'presympathetic' and neutral stance. But if there is from the start no sympathy, how can the other know what a 'yes' or a 'no' sign for sympathy looks like and how neutrality is expressed? Indeed, how can I know it myself? More abstract notions (e.g., 'positive attitude') face the same problem. (At issue is, of course, a version of the classical problem of subjectivity.)

If there is initially no sympathy between me and the other, how could I have any notion of what 'positive' means? Someone might say, 'By trial and error'. But if I have no notion of sympathy, how am I to identify a certain response as 'positive'? There is no room for trial and error, for I have no notion of what to make of the responses. In fact, I could not even know that what I take to be responses are responses. Still further, one can wonder how I can have any notion of either sympathy or responding at someone else.[7] (Given the current debate in evolutionary psychology, there seems to be reason to stress the fact that 'innateness' does nothing to solve this problem.)

My general point with this reflection is that we cannot arrive at understanding each other by way of inference; we have to – and do – understand each other from the start. With the infant, this understanding is limited but it is there. (For more on this, see Hobson, 2003.) My central point is that understanding cannot get off the ground without sympathy. Showing that understanding cannot be merely an intellectual activity where different signs are given hypothetical interpretations that are then tested points precisely to the sense in which sympathy is an irreducible aspect of understanding. This is, in my view, also Wittgenstein's point when he, discussing the possibility of beings that, although intelligent, would be utterly unintelligible to us, says: 'What is essential for us is, after all, spontaneous agreement, spontaneous sympathy' (Wittgenstein, 1998, § 699).

It is important to see that *in so far as* sympathy involves understanding it *ipso facto* involves a desire for, and wish for, the well-being of the other. In fact, *concern* for the well-being of others can only be understood in terms of sympathy or, in other words, in terms of a desire for the other as such. For if this desire was lacking, one could not be *concerned about* the well-being of others; that is, one could have no conception of the other's well-being.[8]

Here one could point out a conceptual issue that possibly had significance for Westermarck. In Swedish, which was Westermarck's native language, there is the word *lust* (etymologically related to English 'lust'), which means at once both desire for the other and joy over being with the other. If we think of sympathy along the lines of lust, it becomes clear that sympathy cannot be simply an urge for the other. As we shall see, Westermarck seems to think of sympathy along these lines even if he fails to see the full consequences of his own thinking on this issue.

Being concerned about the other's well-being is a concern 'as such' irrespective of one's own well-being. The wish that goes with sympathy thus has a moral character and is therefore, to put it in Kantian and problematic terms, both categorical and absolute – a fact that Kant failed to see.

Kant took it for granted that all feelings, with the exception of the respect for the moral law, were simply self-regarding bodily urges. He did not see how a closer analysis of sympathy falsifies his account of sympathy or love. Nor did he see that given his account of feeling, he cannot make sense of moral feeling.

For Kant, feeling lacks understanding. We can use our reason in order to better satisfy the cravings connected to feelings, and we may take it as our moral task to help others satisfy their bodily cravings, but such cravings cannot be the determining principle of moral understanding because they do not contain any understanding. This is the basic dualism in Kant's philosophy, and evolutionary psychology falls prey to the same dualism with the difference that it rejects the idea of a rational moral law and accepts only what Kant calls theoretical reason. Evolutionary psychologists fail, however, to see something that Kant did see: that for theoretical reason there is no such thing as morality.

Daniel Dennett completely fails to see that in terms of theoretical reasoning or, as I would say, any kind of reasoning, it is impossible to form any conception of freedom and morality. In addition to that Dennett is, like Kant, completely entangled in the metaphysical opposition between altruism and egoism (see Dennett, 2004, pp. 194–7). He is also committed to the popular idea that moral understanding is a social phenomenon (see esp. Dennett, 2004, pp. 169 ff.).[9] In addition to this Dennett has adopted Kant's erroneous account of feelings.[10] Thus, Dennett is left with most of the problems in Kant's view, and he has missed the Kantian insight about (theoretical) reason and morality.

What I have tried to show is that sympathy is not a blind feeling of pleasure, but a way of understanding and feeling other persons in the light of love. I will not go into the question whether and to what extent feelings other than sympathy can be understood as merely blind urges. Here it suffices for me to note that understanding constitutes an important aspect of sympathy. I will now discuss Westermarck's theory of morality and the way it attempts to place sympathy at the heart of morality.

Impartiality and sympathy

The core of Westermarck's moral theory is made up of a dynamic between two main elements: (1) emotions between individuals, and (2) socially produced emotions that Westermarck characterises as 'disinterested'. The most developed 'moral' level, the level where *impartiality* is present, is a result of a dynamic between individual and social emotions. Within the individual emotions he distinguishes, among other things, between associative sympathy and altruistic sentiment or affection, also characterised as sympathy in the popular sense (ER, pp. 96–7). Disinterested retributive emotions are produced by 'cooperation' between these two emotions. Associative sympathy is involved when we associate feelings with their outward expressions: 'a happy face tends to produce some degree of pleasure in him who sees it' (ER, p. 96), whereas affection is at stake when we have a 'conative disposition to promote the welfare of its object' (ER, p. 97). The way Westermarck describes the latter feeling, it has important points of contact with the way I have described sympathy earlier, namely as a feeling that is not concerned with the agent's own self but with the other.

Affection for others becomes disinterested by the workings of associative sympathy. In society there is 'mutual good-will, harmony and sense of solidarity' and so we tend to react with disinterested kindliness to people and we react with disinterested resentment if the members of society are hurt (ER, p. 104). Facts such as these led Westermarck to say that the solution to the problem of impartiality

> is not difficult to find. It lies in the fact that society is the birth-place of the moral consciousness; that the first moral judgements expressed, not the private emotions of isolated individuals, but emotions felt by the society at large; that tribal custom was the earliest rule of duty.
>
> (ER, p. 109; see also p. 50)

That is: the affection we feel for each other is generalised; it becomes disinterested because of our inclination to associate with the pains and pleasures of the members of society. When this generalised feeling becomes strong enough, it gives rise to the idea of impartiality, to the feeling that what is morally right is right equally for all. Westermarck is careful to point out the conceptual aspects of this issue. What matters is the *idea* that justice means equal treatment of all – not how far impartiality is actually implemented in different cases. In this way customs – common norms for behaviour and 'the earliest rule of duty' – arise together with the generalised feeling. Breaking norms arouses resentment among people in a society (ER, pp. 93, 108–13).

Westermarck *seems* to think that sympathy is a feeling that must be understood in its own right. He also thinks that Kant is wrong in thinking that moral feeling has its source in intellectual insight. Instead, he says, it is moral feelings – for there are a number of them – that give rise to intellectual ideas about morality. However, Westermarck also thinks that morality evolves as a result of the evolution of impartiality and intelligence (see pp. 69 and 89 ff.). These latter ideas seem to be at odds with the first mentioned ones. If intelligence is needed for sympathy to arise, then Westermarck would have to show how intelligence could be internally related to sympathy, without this inviting the problems Kant grappled with.

Westermarck's difficulties are in a sense the mirror image of Kant's difficulties. Westermarck approaches the problem of morality from the direction of feeling, but, without recognising this, ends up saying that some form of reason is indispensable to morality. With Kant things are reversed. The relation between Westermarck's and Kant's accounts serves to show how philosophical oppositions share the same conceptual space. This means that in a deeper sense they share the same outlook.

Westermarck differs from Kant in that in his account, sympathy seems to contain an inherently moral feature of the kind I described earlier. However, he is unable to connect his account of sympathy with his view of morality. Failing to see the sense in which sympathy is a concern for the other as such, he brings in the concept of impartiality. This is the move that brings him so close to Kant.

Westermarck's inability to incorporate sympathy into his account becomes evident in that he says that only generalised, impartial feelings can count as moral. Thus, it is not the affection that people have for each other that is moral. The only *moral* feeling there is addresses no one in particular and it is only on *this* account that it is taken to be moral. However, this way of conceiving morality is essentially Kantian (cf. Kant, 1996, pp. 76 ff. or AA5:57 ff.).

It is quite clear that Westermarck's reason for assuming that impartiality constitutes a criterion for morality is that he thinks that in want of impartiality, the feeling of affection is biased in a selfish way. This is why he emphasises altruism, disinterestedness and impartiality. Pursuing purely personal interests and desires is amoral, and desires become moral only to the extent that they lack personal interest. This is taken to mean the same as having a general interest.

Westermarck introduces impartiality for exactly the same reason as Kant appeals to the universality of reason, namely in order to steer clear of the selfishness that 'subjective' feelings have. What we have is in other words a typical metaphysical constellation: objectivity is supposed to save us from falling into an inscrutable subjectivity. And when things are understood from the point of view of objectivity, the difference between subjective feeling (whatever that is) and sympathy between persons *cannot* be seen. Thus, the latter is interpreted as an instance of the former. In the case of Westermarck this means that the account he gives of sympathy is in the end completely ignored, and an objective account of 'impartiality' becomes criterional for morality. In the end, morality is all about the *idea* of impartiality that has its source in 'emotions felt by the society at large' (ER, p. 109).

At this stage we have to pose similar questions to Westermarck that we posed to Kant: If the feelings of individuals are assumed to be inherently selfish, what moral role can they have at all? And how could there be any such thing as sympathy? In the end the sympathy that individuals feel for each other has no moral role, whereas the element that is assumed to have a crucial moral role has a distinctly Kantian ring to it.

How can a nonmoral affection be altruistic? How can Westermarck describe this sentiment as a 'conative disposition to promote the welfare of its object' (ER, p. 97) while at the same time thinking that it is not a moral disposition? The reason for this oddity in Westermarck's thought is that he is too much of a Kantian. Like Kant, he assumes that benevolent feelings are not, as such, moral. They 'become' moral only when they can take the form of impartial judgements, that is, when they have acquired a general, conceptual dimension. When we remind ourselves of how much Kant tried to stretch his thought in the direction of emotion, declaring even that a human being who would lack moral feeling would be 'morally dead' (Kant, 2000, p. 160 or AA6:400), we realise that the difference between him and Westermarck is not important.

Nevertheless, Westermarck in fact notices a crucially important aspect of sympathy and the sense in which it is a moral feeling. If I am allowed a little guessing, I think that it is this insight that forms the core of his moral view. The problem is that he does not see the philosophical potential of his insight. In making the typical move of philosophers and scientists, namely trying to give an objectively valid account of morality, he loses contact with his insight and instead falls into the problematic perspective into which not only Kant but philosophers and evolutionary psychologists in general have tended to fall into.

Westermarck perhaps had a glimpse of the fact that sympathy is at odds not only with Kantianism, but with the idea of giving an objectively valid account of morality. Because he was unable to see the full meaning and significance of sympathy, he misinterpreted his insight as providing a proof for relativism. As a result, his main objection to Kant is not about the meaning of morality but about the truth of moral claims. When it comes to meaning, Westermarck's account does not, as we have seen, differ from Kant's: morality is taken to be founded by the idea of impartiality, the difference being that Westermarck thought that such ideas cannot be true a priori. This is 'because' they have their origin in collective feelings.

What Westermarck does not notice is that the meaning of impartiality the way he himself describes it is absolute in character. The moral point of impartiality lies precisely in the idea that *all* persons *must always* be treated *equally*. Westermarck explicitly says that the common feeling of impartiality cannot be 'knowingly partial' (ER, p. 93). This, however, actually means that it is the concept of moral rightness that gives sense to impartiality. Westermarck ends up saying that if people lack the *conception* that in certain cases equal treatment of all is what it *means* to act morally, then they lack moral understanding. But this amounts to saying that *conceptuality*, more particularly the concept of impartiality, is criterional for moral responses. What Westermarck says leads to the view that morality involves a certain way of *thinking* about moral acting. Insofar as we take morality seriously, this means that we cannot, on Westermarck's terms, help thinking in terms of impartiality. Does this not mean that impartiality is like a moral law? It secures morality by virtue of its universality.

Conclusion: philosophy and sympathy

Ironically, both Kant and Westermarck ended up giving a decisive role to that very 'element' that they saw as their main task to resist. The obscurity that causes this oddity is inherited by contemporary evolutionary psychology. Evolutionary psychology cannot account for sympathy because sympathy is all about a concern for the other, not for one's own well-being or the well-being of the community. Evolutionary theories cannot give an adequate account of sympathy because they assume that if it is not in itself selfish, it must be beneficial on the group level. Some theorists explicitly claim that 'pure' unselfish sympathy is not compatible with evolutionary theory (van der Weele, 2011). They assume that only feelings that have an adaptive value 'can' exist. Because sympathy does not have any adaptive value, it cannot exist. For instance, van der Weele clearly implies that sympathy is just an illusion. It is characteristic of evolutionary psychologists that they easily make frivolous and thoughtless assumptions of this kind. Ruse and Wilson (1986, pp. 178–86) even claim that morality as such is an illusion created by the genes. We 'function' better under this illusion.

But any conception of the 'functioning better' of human beings is moral. Mother Nature does not care a whit whether creatures like *Homo sapiens* exist. It is confusing to think of Mother Nature or natural selection as something that cares.[11] Only individual beings can care. That we humans do care is an aspect of our moral understanding. The further extensions of this understanding can be studied only in terms of morality, *not* in terms of the mechanics of nature. When efforts of the latter kind are actually made, it sounds like this: 'So value exists for animals solely because natural selection built neurocomputational circuitry into our minds to compute it as one of several kinds of representation necessary for regulating our behavior according to evolutionarily functional performance criteria' (Tooby, Cosmides, & Barrett, 2005, p. 316). Thus, the computational architecture 'must know (in some sense) that living children are better than dead children, social approval is better than disapproval, salt and sweet are better than acrid or putrefying [. . .], and so on' (Tooby, Cosmides, & Barrett, 2005, p. 317).

But an egocentric logic could decide only whether it is better to eat a child or to use it for labour, and such logic has got *nothing* to do with morality – for morality is all about sympathy, about caring for the other as such. Given that we have this understanding, computing the 'value' of a child is a horrible undertaking.

Notes

1 'Sympathy' is far from being an adequate word for what I try to say (in an important sense 'love' would be better), but given the aim of this chapter, 'sympathy' probably causes the fewest problems.

2 It would in fact be better to say that sympathy is not one of the feelings (emotions, affects or moods). I cannot go deeper into this question, but what I say does indicate what I mean.
3 In fact, and on Kant's own terms, because theoretical reason cannot recognise freedom and because practical reason has to presuppose it, the whole question of determinism and indeterminism is a misconception. (For some contemporary standard views on the issue, see Dennett, 2004; McCann, 2006; Meixner, 2006).
4 Taylor tries to show how self-esteem can 'encourage' one 'to see others as valuable' (p. 272). But if one were to lack sympathy, it would all be a form of self-regard. And if one has sympathy, one already sees the other as valuable, although from the point of view of sympathy, this way of putting it is inadequate.
5 For more on this, see Backström (2007, pp. 91 ff).
6 Here one could perhaps apply Richard Gaskin's point that having a certain feeling does not involve being able to give a conceptual account of it. See Gaskin (2006, p. 148).
7 Cf. Wittgenstein: 'If my personal experience is all I know how can I even assume that there is any other besides?' (Wittgenstein, 1999, p. 261).
8 Although I am in agreement with much of what Evan Thompson (2001) says about empathy, I also disagree with him on some very central issues. First Thompson's view is unable to account for the lure of collectivity (see fn. 9). Second, although Thompson does say that interpersonal understanding is impossible without empathy, his account is still governed by metaphysical concerns rather than by the meaning of the concern between 'I' and 'you'.
9 I have elsewhere shown that morality is not a collective affair. The apparent plausibility of the idea depends on confusing conscience with something I have called collective pressure. This confusion is characteristic not only of Kant and Freud but of philosophers and psychologists in general (see Backström & Nykänen, 2015, forthcoming). Westermarck's theory is hampered by the same confusion (see ER p. 54).
10 I cannot refrain from pointing out that Dennett's account of Kant's ethics is typically and fundamentally flawed: 'The Kantian ideal is a fantasy in which you somehow strengthen your pure-reasoning muscle to such a fine pitch that you can make pure, emotionless judgments untainted by tawdry guilt feelings or base longings for love and acceptance' (Dennett, 2004, p. 213). If Dennett had cared to take a little closer look at Kant's writings, he might have realised how off target his account is – and how evolutionary psychologists have generally taken over Kant's account of feelings.
11 McKay's and Dennett's paper 'The evolution of misbelief' goes amiss because of, among others, this confusion. They speak of moral concerns as if there were some criteria external to morality and set by nature, criteria that can decide the utility of moral acting and thinking. See McKay & Dennett (2009, esp. p. 507).

16 Darwinian conservatives and Westermarck's ethics

A political dimension of the late twentieth-century Westermarckian renaissance

Antti Lepistö

Introduction

It is not uncommon to see Westermarck cited as 'the first sociobiologist', 'the first Darwinian sociologist', or as a pioneering scholar who proposed 'arguably one of the first evolutionary psychological hypotheses', that is, the Westermarck hypothesis concerning incest avoidance (Fessler, 2007, p. 355; Sanderson, 2006, p. 24; 2007, p. 74). What is perhaps less well known even among Westermarck scholars is that his evolutionary ideas have been invoked also in the context of some of the most heated *political* debates of the late twentieth and early twenty-first centuries.

This chapter offers an analysis of a specific political dimension of the current Westermarckian 'renaissance', namely the use of his theories of morality and the family by American conservative intellectuals since the late 1980s. Recent American right-wing political commentaries on Westermarck have typically been contributions to two different but related ideological issues: first, the debate on the 'decline' of the traditional two-parent family, and second, the debate concerning the role of the civil society as opposed to the federal government. Four conservative thinkers are of particular interest here: leading neoconservative political scientists James Q. Wilson (1931–2012) and Francis Fukuyama (b.1952), self-described 'Darwinian conservative' Larry Arnhart (b.1949) and journalist and editor of the paleoconservative magazine *Chronicles* Thomas Fleming (b.1945). Although it is possible to find occasional references to Westermarck in other texts published by U.S. conservatives, these four are, to the best of my knowledge, the ones who most consistently integrate Westermarck into their political and/or scientific argument (for others, see, e.g., Blankenhorn, 2009, pp. 37–8, 94; apart from the conservative academic and journalistic circles discussed in this chapter, Westermarck's work on the family has also been used recently in the U.S. Supreme Court by religious groups to oppose same-sex marriage: see the amicus curiae brief of Evangelical and other religious organizations in *Hollingsworth v. Perry* (2013, pp. 8–9). It is above all Westermarck's evolutionary reasoning that has inspired the conservative intellectuals. This is significant considering the frequent assumption that conservatives in the United States as a rule detest Darwinism and evolutionary science. To understand the recent right-wing enthusiasm for Westermarck as a Darwinian thinker, it is essential to see that, in the last forty years, an important subgenre of intellectual conservatism known as 'Darwinian conservatism' or 'neo-Darwinian conservatism' has emerged. (It should be pointed out that this study is an offshoot of my larger research project on recent moral and cultural conservatism in the United States and that an elucidation of Westermarck's significance for liberal political thought would require another study.)

Conservative intellectuals, evolutionism and Westermarck

'Neo-Darwinian conservatism' in the United States was recently examined by Jon Anstein Olsen, who traces the origins of the neo-Darwinian conservative project – or, as he emphasises, the different projects – to the 1970s and introduces the conservative writer George Gilder as the most important pioneer of the movement (Gilder, 1973; Olsen, 2013, pp. 129–33). However, Olsen (2013, pp. 7, 125, 283) is certainly right to note that Neo-Darwinian conservatism emerged as a more widespread, overt enterprise only in the late 1990s and early 2000s. By then, it was Larry Arnhart who put in the most effort to make the 'broader, explicit case that Darwinism supports conservative ideology' (Olsen, 2013, p. 93). Olsen in fact discusses each of the four thinkers identified in this chapter as Westermarckian conservatives – all of whom have been mainly interested in 'human nature' rather than in genetic differences.

In the following, I will argue that the four conservatives Wilson, Arnhart, Fukuyama and Fleming were drawn to Westermarck's work because of three themes in the Finn's oeuvre. They were (1) maternal and paternal emotions, (2) (global) mating patterns and the emotions that guide them and (3) the incest taboo. Although the topics in themselves suggest that there were serious scientific issues under consideration, a mere glance at the titles of the 'Westermarckian' conservative works is enough to indicate extra-theoretical interests. *The Politics of Human Nature* (Fleming, 1988), *The Great Disruption: Human Nature and The Reconstitution of Social Order* (Fukuyama, 1999), *The Marriage Problem: How Our Culture Has Weakened Families* (Wilson, 2002) and *Darwinian Conservatism* (Arnhart, 2005b) were works that promoted a more or less open political agenda. By and large, the authors shared the convictions typical of American moral or cultural conservatives of the era: for example, they voiced misgivings about liberal individualism, hostility to moral relativism and a high regard for virtue ethics. However, their main political interest lay in two specific, major late-twentieth-century American debates: the decline of the family and the role of the civil society as opposed to the state.

Westermarck and the war over the family

In post–Second World War United States, the case for the strengthening of the nuclear family was not only a conservative issue. As the historian Robert O. Self notes, the white, middle-class, nuclear family ruled by a patriotic and heterosexual male had been the focus of liberal concern since the New Deal, and that way it continued in the 1960s. At the time, 'New Frontier and Great Society liberals such as Lyndon Johnson, Daniel Patrick Moynihan, and Sargent Shriver crafted social and economic policies they believed would make this idealized nuclear family . . . attainable for more Americans than ever before'. The collective effort by liberals to assist families economically is labelled by Self (2012, p. 4) as 'breadwinner liberalism'.

According to Self, things began to change only after the mid-1960s, when feminists, gay rights activists and antiwar demonstrators began to contest post-war Americans' cultural assumptions about gender roles, sexuality and masculinity. These challenges in turn provoked American conservatives in the following decades into a forceful reaction (yet again) built around an idealised conception of the nuclear family. The result was, in Self's terms, the emergence of 'breadwinner conservatism' which 'endeavoured to defend families from what they cast as moral threats'. Over the four decades from the early 1960s to the early 2000s, 'conservatives won back the power to define the mythological American family, and in the process the Great Society pledge to assist families became the New Right pledge to protect them' (Self, 2012, pp. 4–5).

Whereas Self underlines the evolution from breadwinner liberalism to breadwinner conservatism, thus highlighting the nuclear family as a conservative emblem in recent American politics, other scholars have emphasised the significance of the family issue for all political factions. For example, the political scientist Robert Booth Fowler (1999, p. 214) argues that the endeavour to further the cause of families expanded during the 1980s and 1990s into an entire 'family revival movement' which crossed all conventional political boundaries. Fowler (1999, pp. 215–20) is quick to add, however, that the many bones of contention pertaining to family – such as divorce, single mothers and above all children – justify the expression 'war over the family', despite wide-ranging support for the family movement as such.

Omitting the question as to what extent family revival has been on the agenda of liberal or leftist political thinkers and agents in recent years, it is evident that for conservatives, the issue has been critical. All four conservative Westermarckians mentioned earlier have visibly participated in the pro-family campaign. Thus they have bemoaned the current condition of American families, criticised liberal and leftist views of family and marriage, and in various ways stood up for the traditional two-parent family. Importantly, three of them – Wilson, Arnhart and Fleming – have cited Westermarck's research in the context of these debates. The neoconservative James Q. Wilson, who has been described as 'the most sophisticated intellectual proponent of the conservative crime response program' initiated in the 1970s, is clearly the best known of the three (Bowman, 2006; Ruth & Reitz, 2003, p. 81).[1] Importantly for our present purposes, Wilson cited Westermarck's research – as many as five different works: ODMI, HHM, SHHM, ER, *Marriage* (1929) – in two of his books, first in *The Moral Sense* in 1993 and later in *The Marriage Problem* in 2002. On both occasions, the context was Wilson's discussion of the family and its role in the maintenance of social order.

In *The Moral Sense*, Wilson's key argument was that the human mind as a product of evolution by natural selection is predisposed both to make moral judgements and to acquire 'a set of social habits that we ordinarily find pleasing in others and satisfying when we practice them ourselves' (Wilson, 1993, p. 25). Importantly, however, the development of these universal or 'natural' dispositions – which Wilson saw as the (psychological) foundation for 'non-criminality' in most people – was not entirely spontaneous but required a domestic environment in the individual's early childhood that was not too different from the average caring family. As Wilson (1993, p. 2) himself put it, the moral sense was 'formed out of the interaction of [children's] innate dispositions with their earliest familial experiences'.

Wilson's argument for the naturalness and universality of the moral sense, then, rested on the idea that there were 'fundamental social bonds', in particular, 'the practice of child nurturance', that could be regarded as human universals (Wilson, 1993, p. 18). It was here that Westermarck's studies came into play. In Wilson's assessment, Westermarck had argued very convincingly for the universality and naturalness of the nonrational emotional attachments between family members, and therefore his studies could be used to counter the 'oddly' influential view that 'love of children' was 'a modern invention' (Wilson, 1993, p. 19). In the early pages of *The Moral Sense* Wilson thus noted how

> Westermarck's magisterial two-volume account of the origins of moral ideas among primitive peoples concluded that "the maternal sentiment is universal in mankind" and that "parental affection is not a late product of civilization, but a normal feature of the savage mind".
>
> (Westermarck, 1906a, pp. 529–31; Wilson, 1993, p. 19)

What most concerns us here is how Wilson used Westermarck's empirical assumptions about human familial emotions to highlight his *political* argument about a threat hanging over American society, namely the 'decline' of the traditional two-parent family. In Wilson's estimate, 'the prevalence of divorce and illegitimacy' (Wilson, 1993, p. x) were clear signs of a weakening of the family institution in America that could have harmful effects on the development of the moral sense in children, thus leading to social problems such as the increase of crime. Wilson's intention was to highlight both the seriousness and the unnaturalness of the situation by referring to the empirical facts that Westermarck had presented concerning familial and moral emotions. To underpin his conservative cultural criticism even further, Wilson (Wilson, 1993, pp. 148) cited also recent research which complemented Westermarck's earlier findings on parental affection, showing that the moral sense or moral sentiments were more easily developed in 'securely attached children' than in 'insecurely attached ones'. Then taking up his role as a cultural critic again, Wilson argued that, in the current American society, the parent's natural predisposition to attachment towards the child operated in extremely difficult conditions. In modern America, adults lacked economic incentives for having children, with the result that 'fewer children will be produced, marriage (and childbearing) will be postponed, and more marriages will end in divorce'. What is more, many children were raised in ways that 'minimize the parental cost in personal freedom, extra income, or career opportunities' (Wilson, 1993, p. 159). In such conditions it was vital, Wilson thought, to consciously cultivate the evolved natural family bonds. Only thus would it be possible to protect and develop the natural moral sense, 'our persistent but fragile disposition to make moral judgments' (Wilson, 1993, p. xv).

Wilson returned to Westermarck later in *The Marriage Problem: How Our Culture Has Weakened Families* (2002). This time Wilson argued even more explicitly than he had done in 1993 that in the United States, the institution of marriage and therefore also the family, 'the foundation of public life' (Wilson, 2002, p. 7), were encountering serious problems. He was particularly unsympathetic to the 'countless' critics of marriage and the two-parent family, who assessed their significance in relation to their contribution to some (sexist) 'traditional culture', and who thus highlighted their role as a source of female unhappiness. Although admitting that 'there is merit in this criticism [of the 'traditional family'] that cannot be wished away' (Wilson, 2002, p. 13–14), Wilson wanted to replace such an analysis with one that understood the family as a universal institution and as a matter of biological dispositions interacting with cultural factors. Once again he cited Westermarck to remind his readers of the biological aspects of the question.

In brief, Westermarck's studies – more precisely, his extensive comparative studies of human marriage patterns and mating in the animal world – were important for Wilson because they showed monogamy to be the most common type of marriage and polygamy to be an exception. They provided support to the idea that biology, and not only culture, was 'dictating marriage patterns' (Wilson, 2002, pp. 46–7). Wilson had to concede, however, that the case was rather complicated. Although evolution apparently had 'selected for sentiments that encourage marriage', these familial emotions were not of the kind that would *require* marriage (Wilson, 2002, p. 105). In the end, Wilson (Wilson, 2002, pp. 24–5) admitted, biology 'does not go very far in explaining why a father is part of the family', even if it was true that 'nowhere do we find a place where children are regularly raised by a mother who has no claims on the father'.

Turning next to the impact of culture and taking up his role as a cultural critic, Wilson found the problem with contemporary American culture to lie in the excessive respect for individual rights, stemming from Enlightenment ideals: culture had made *cohabitation* – characterised by lack of ultimate responsibilities towards the other – a lifestyle favoured by the young (Wilson, 2002, pp. 3–7, 18–21). In addition, in African-American communities families were still encumbered with the legacy of slavery, which had taught mothers and children not to expect the presence of the father, and the reason why boys lacked a model of the father as a (proper) husband (Wilson, 2002, pp. 126–9). Wilson's (Wilson, 2002, p. 14) key message was, therefore, that 'the one indispensable link between family and culture is that the latter supports the former'. If the United States wished to continue as a vital, thriving nation, the country would have to bring into being a culture that supported more effectively than before the Westermarckian 'sentiments that encourage marriage' and the ultimately fragile institution of marriage.

Larry Arnhart relies even more heavily on Westermarck in his enquiries on the family. Arnhart, a political theorist and the leading proponent of Darwinian conservatism in contemporary America, typically approaches questions relating to family from the perspective of the history of political philosophy, assessing the strengths and weaknesses of classical political texts on the basis of their consistency with modern scientific views. Despite this background in philosophical scholarship, he does not hide his political beliefs. Arnhart has considered himself a 'liberal conservative' or a 'fusionist' who brings together ideas from both traditionalist and libertarian conservatism. Essential to his type of conservatism is the separation of state and society and a focus on the role of (civil) society, including the family, in the construction and preservation of moral order (e.g., Arnhart, 2006).

Arnhart has been very explicit about his intellectual debt to Westermarck. For example, in a lecture on Westermarck in Helsinki, Finland, he noted that his discovery of Westermarck's importance in the late 1990s helped him articulate his long-term view that Darwinian research supports the revival of Aristotelian ethical naturalism (Arnhart, 1998a, p. 1).[2] The general idea of Arnhart's program – which he calls 'Darwinian natural right' – is that

> the good is the desirable, and [that] therefore Darwinian biology supports a science of ethics by explaining natural desires that are universal to all human societies because they express a universal human nature. The satisfaction of these natural desires constitutes a universal standard for judging social practice as either fulfilling or frustrating human nature, although prudence is required in judging what is best for particular circumstances.
>
> (Arnhart, 1998a, p. 1; see also Arnhart, 1998b)

In the context of his 'Darwinian natural right' philosophy, Arnhart has discussed several Westermarckian ideas relating to the family. A central concept here is the idea that natural human 'desires for family life' (Arnhart's term) are based on instincts and moral emotions that evolved by natural selection. These 'Westermarckian' familial instincts and emotions include, for example, the instincts for parental care and the inclination to feel moral indignation toward men who abandon their wives and children (Arnhart, 2005a, p. 199, 2011).

Remarkably, Arnhart not only uses Westermarck's empirical observations on familial emotions to lay the ground for his own argument, but he also insists that Westermarck himself was an early proponent of an analogous natural right theory. He underlines the fact that Westermarck could foretell, on the basis of his observations on human nature, the painful consequences that would follow from suppressing the aforementioned familial desires (Arnhart, 1998a, pp. 1, 7). According to Arnhart (Arnhart, 1998a, p. 6), Westermarck therefore 'clearly relies on the uniformity of human nature as a ground for universal ethical principles'. Thus he understands Westermarck's position to be 'essentially the same' as his own 'Darwinian natural right', in which the satisfaction of 'natural desires' forms the basis for universal standards for judging social practices (Arnhart, 1998a, p. 1). Such a reading of Westermarck's ethical theory, of course, seems to collide with Westermarck's explicit embrace of ethical relativism. I will return to this question.

In addition to the more general notion of defending 'family life', Arnhart's Darwinian natural right philosophy advocates marriage as a 'permanent moral standard' (e.g., Arnhart, 2007). In this context, Arnhart draws on Westermarck's studies both on mating patterns in human cultures and in the animal world and on the instincts or emotions that guide these patterns. Arnhart credits Westermarck, first of all, for his Darwinian notion of marital bonding, according to which 'conjugal bonding has always been a natural instinct for human beings, because it was favored by natural selection to provide parental care for offspring that could not survive without such care' (Arnhart, 2005a, p. 199). Second, Arnhart notes approvingly Westermarck's keen anthropological eye for cultural 'regularities that manifest the universal nature of human marital emotions'. Importantly for Arnhart, these 'regularities' were the same as those seen by St Thomas Aquinas, Adam Smith and Charles Darwin (Arnhart, 2001, pp. 20–1, 2005a, pp. 199–200). The question here is about the different forms of marriage and their global distribution, and importantly for our study, Westermarck is credited for presenting not only anthropological data, but also a plausible explanation for his own observations – and for those by Aquinas and others. The key to his explanation was jealousy and its role in setting the limits for marital relations.

To be more specific, Arnhart credits Westermarck for confirming Aquinas's and Smith's analyses on polyandry and polygyny by showing, first of all, that polyandry was the rarest (and most unnatural) form of marriage, 'because the intense jealousy of men makes it almost impossible for them to share a wife' (Arnhart, 2001, pp. 20–1, 2005a, p. 200; see also Westermarck, 1908, pp. 387–92, 1921c, pp. 104, 107, 206, 221–2). Polygyny was more common, because many men had a natural desire for multiple mates. However, in this form of polygamy wives had a natural inclination to sexual jealousy, which created 'conflicts that are difficult to manage' (Arnhart, 2001, pp. 20–1; 2005a, p. 200). The most common – that is, universal – form of marriage was monogamy 'because it satisfies the human instincts for conjugal bonding and parental care' (Arnhart, 2001, p. 20, 2005a, p. 200). For Arnhart, Westermarck's observations therefore represent evidence for two related claims: that 'human marriage is natural' and that 'monogamy is the most fully natural form of mating' (Arnhart, 1998b, pp. 262–5, 2001, p. 20, 2005a, p. 199). Although Arnhart, as noted, often discusses Westermarck in a philosophical rather than in an urgent political context, he has also used these Westermarckian ideas in a more straightforward, political attack on the left, using them to criticise Marx's and Engels' notions of the family as an institution that will be abolished by the communist revolution (Arnhart, 2005b, pp. 48–9). It is not unreasonable, therefore, to read Arnhart, like Wilson, as one of the authors who mobilised Westermarck's family theory for the late-twentieth-century American 'war over the family'.

Lastly, Thomas Fleming, the classics PhD and journalist, is another conservative writer with an interest in Westermarck's Darwinian studies of the family. Admittedly, Fleming cites Westermarck only in passing in the context of a more general discussion on the incest taboo; nevertheless, his book *The Politics of Human Nature* (1988) should remind us that not only moderate conservatives such as Wilson and Arnhart, but also reactionary right-wing thinkers have found use for Westermarck in their political treatises. Fleming has been known since the early 1980s as an intellectual leader of the 'paleoconservative' movement, and, since the mid-1980s, he has worked as the editor of *Chronicles: A Magazine of American Culture*. The journal – as well as the entire paleoconservative movement built around it – is known for its opposition to immigration, interventionism, globalism and neoconservatism (which is found to be too liberal and universalist) (Fleming, n.d.; Francis, 2002). Fleming's *The Politics of Human Nature* was, above all, a case for patriarchal authority, family autonomy and states' rights, one that drew heavily on Darwinism and sociobiology.

For us, Fleming's book is of interest because his notions of the 'natural family' (Fleming, 1988, p. 113) and the good society rested in part on a Westermarckian reading of the incest taboo. Fleming counted the incest taboo among the 'natural laws' that constituted the foundation of flourishing and productive societies. This was, above all, because it 'will tell him [any member of a family] with whom he can mate and have children'. 'In this sense', Fleming (Fleming, 1988, p. 110) argued, 'the incest taboo can be seen to establish the fundamental parameters of all family and social relations'. Importantly, Fleming's reading of the incest taboo was explicitly Westermarckian (as well as anti-Freudian and anti-Durkheimian). Fleming (Fleming, 1988, p. 112) noted how Westermarck had observed the existence of instinctive aversion against sexual relations with cohabiting siblings among many animal species and, 'over sixty years ago, argued that the incest taboo was a natural development of this tendency'. For these reasons, relating to the functions and the natural history of the incest taboo, Fleming (Fleming, 1988, p. 130) counted it as one of the 'fundamental principles of our social life' – along with maternal care, paternal authority and sexual division of labour – which could not be destroyed without wrecking 'the entire social and political edifice erected upon these natural principles'. The targets of Fleming's often ruthless criticism were those ideas, institutions and laws of the modern world that supposedly threatened the functions and roles of the traditional family – in particular the 'maternalistic' welfare state as well as individualist law reforms, including no-fault divorce (Fleming, 1988, pp. 117, 196–9).

Westermarck and civil society theory

Westermarck's work has been used by conservative intellectuals also for other purposes besides promoting the 'traditional family'. Some of the conservative texts citing Westermarck are best understood as contributions to the debates on 'citizenship' and 'civil society' that emerged at the centre of U.S. political debate in the 1980s and 1990s.

As noted by analysts of contemporary political philosophy, this debate developed as a continuation of the important but nearly log-jammed exchanges between liberals and communitarians on individualism and community. At the time, the concept of citizenship became the key to the advancement of American political theory, because it was related to both liberal ideas of individual rights and entitlements and communitarian notions of membership in particular communities and attachment to them (Kymlicka, 2002, p. 284). In addition, observable political events and trends, such as the prolonged dependence of citizens on Social Security and increasing voter apathy, stimulated an interest in citizenship issues. Consequently, many American political theorists on the right, left and centre came to the conclusion that without citizens with particular qualities (virtues), democratic societies are unstable and difficult to govern (Kymlicka, 2002, pp. 284–5; Kymlicka & Norman, 1994, esp. p. 353).

In the context of this debate, so-called civic republicans argued that participation in public life inculcates civic virtue (e.g., Kymlicka & Norman, 1994, pp. 361–2). Many right-wing thinkers, in contrast, claimed that the free market works as a teacher of civic virtue by requiring people to display initiative, self-reliance and self-sufficiency. According to them, the welfare state promoted passivity (among the poor) and created a culture of dependency (Kymlicka & Norman, 1994, pp. 355–6). Still other theorists of civic virtue were in favour of civic education in schools (Kymlicka, 2002, p. 307). Importantly for our concerns, however, there was yet another set of theorists focusing on 'civil society', claiming that virtues of mutual obligation are cultivated by voluntary organizations, including churches, families, trade unions, neighbourhood associations and/or charities (Kymlicka, 2002, p. 305). Although all the aforementioned Westermarckians of the intellectual right would have sympathised with the latter set of thinkers, two of them used Westermarck's ideas explicitly to develop a theory of the civil society: Francis Fukuyama and Larry Arnhart.

In this context, Westermarck's theory of incest aversion received most of the attention. Importantly, Fukuyama and Arnhart saw Westermarck's research on the subject – and later studies that confirm Westermarck's position – as evidence for the natural human tendency to create moral norms within the confines of the civil society, and thus to maintain moral and social order in the absence of an active, hierarchical state apparatus. Along these lines, as part of their larger theories of the civil society, Fukuyama's and Arnhart's Westermarckian readings of the incest taboo translate into arguments for restricting government intervention.

In Fukuyama's case, the most important text for our purposes is his 1999 book *The Great Disruption* (on Fukuyama's neoconservative background and later dissociation from the movement, see Fukuyama, 2006).[3] In the book, Fukuyama argued that the United States had entered a period of 'Great Disruption', beginning in the 1960s. The signs of this disruption were, according to Fukuyama, the increase of crime, enormous changes in fertility and family structure, declining levels of trust and the triumph of individualism over community (Fukuyama, 1999, Part I, esp. 4–5). Behind these negative social trends were, Fukuyama claimed, the post–Second World War technological revolution and the rise of the information society, which had replaced the older industrial society and at the same time, disrupted its distinctive social and moral order. A key argument of the book was that the United States would most likely solve its problems if the political leaders of the country were brought to understand that sometimes the government has to get 'out of the way of individuals and communities that want to create social order for themselves' (Fukuyama, 1999, p. 275). In fact, Fukuyama claimed optimistically that despite the decades-long decline of 'social capital' in the United States, there was already a new social order emerging in the country in the late 1990s.[4] He noted that in the last few years, crime and divorce rates, for example, had actually begun to decline and levels of trust to increase (Fukuyama, 1999, pp. 7–8, 271) and argued that no one should be surprised about such a remaking of social order. As Fukuyama (Fukuyama, 1999, p. 6) put it, 'We can expect this to happen for a simple reason: human beings are *by nature* social creatures, whose most basic drives and instincts lead them to create moral rules that bind themselves together into communities'.

As a validation for his argument, Fukuyama highlighted research in modern biology and economics on 'self-organization' and 'spontaneous order'. Here he also referred to Westermarck's theory of the incest taboo. In brief, Westermarck's theory was, according to Fukuyama, 'one of the best illustrations of how natural instincts can shape social norms in a rather direct way'. Indeed, the incest taboo was 'a good illustration' of norms that evolve 'arationally' and 'spontaneously', in other words, without hierarchical organizations such as the state (Fukuyama, 1999, pp. 159–60).

Fukuyama's book is not to be read as a justification of strict libertarianism, because he recommended also some government action to reconstruct social order (relating, for example, to education and community policing), and he explicitly distanced himself from libertarians (Fukuyama, 1999, pp. 274, 218). However, he recommended – in a moderate fashion – 'stopping the state from doing counterproductive things' (e.g., subsidising illegitimacy) (Fukuyama, 1999, p. 275). To support this argument, he cited studies showing that human beings were biologically hardwired to create moral values, and due to this and other species-typical characteristics, capable to reason 'their way over the long run to spontaneous order' (see the front flap of the dust jacket of Fukuyama's book). Because Westermarck's argument on the incest taboo was cited in such a context, we must conclude that in Fukuyama's book, Westermarck's idea was used to rationalise a moderate right-wing argument in favour of restricting government intervention.

Whereas Fukuyama's 1999 argument for limited government contained many qualifications, Larry Arnhart has been much more systematic in his case for small government. Importantly, he also draws on Westermarck's ideas about incest aversion, and more generally, the origin of moral rules to make his case. We must immediately note, however, that in his only major *political* text (by 2015), that is, his *Darwinian Conservatism* from 2005, Arnhart does not cite Westermarck's incest theory. My argument is that in *Darwinian Conservatism* Westermarck's influence is more implicit than explicit. That is to say, Westermarck's ideas influenced Arnhart's writing in *Darwinian Conservatism* not only where Westermarck is explicitly cited (i.e., relating to family and property), but also pertaining to the central intent of the book – that is, where Arnhart attempts to convince his fellow conservatives of the value of Darwinian science by showing 'how the human capacity for spontaneous order arises from social instincts and a moral sense shaped by natural selection' (see the back cover of Arnhart's book). Looking at Arnhart's analyses of Westermarck's incest taboo theory during the six or seven years prior to the publication of *Darwinian Conservatism* (i.e., 1998–2005), and comparing these to his portrayal of the moral sense tradition in the book itself, we can find a good amount of evidence of Westermarck's implicit influence.

First of all, between 1998 and 2005 Arnhart noted on numerous occasions that he regarded Westermarck as one of the most important 'moral sense' philosophers.[5] Thus in a 2003 article 'Darwinian Conservatism as the New Natural Law' Arnhart cited Westermarck as 'a preeminent example' of a Darwinian social theorist developing 'Darwin's moral sense theory in ways that [are] intellectually and morally defensible' (Arnhart, 2003, p. 15). Second, and more importantly, in the same article, Arnhart's wording to describe Westermarck's moral theory was very similar to what he later used in *Darwinian Conservatism* when introducing his aforementioned main argument on spontaneous order arising from social instincts and a moral sense. To be exact, in his article Arnhart (Arnhart, 2003, p. 16) noted that Westermarck's theory of the incest taboo had been cited by James Q. Wilson, Francis Fukuyama and other (unnamed) conservatives 'as an example of how a Darwinian account of human nature might explain morality and social order as rooted in the moral emotions of human biological nature'. In another article dedicated exclusively to the incest taboo, Arnhart (2005a, p. 203) claimed that Westermarck's notion of the incest taboo 'illustrates how social order arises from the complex interaction of nature and convention: the incest taboo is a social convention that expresses the human nature of the moral emotions'. Considering the fact that *Darwinian Conservatism* was published several years afterwards, we can take it for granted that by then Westermarck's theory of the incest taboo had influenced (or at least confirmed) also Arnhart's own understanding (and not only that of his fellow conservatives) of how 'a Darwinian account of human nature might explain morality and social order as rooted in the moral emotions of human biological nature'.

In *Darwinian Conservatism*, Arnhart used such a reading of the moral sense tradition to revitalise intellectual conservatism in the twenty-first century. Arnhart's most general starting point is the idea that the intellectual vitality of conservatism will depend on the ability of right-wing thinkers 'to show that their position is compatible with' . . . 'new science of human nature' (e.g., genetics, neurobiology, developmental biology and evolutionary theory) (Arnhart, 2005b, p. 3). Closest to Arnhart's heart seems to be, however, the long tradition of moral sense philosophy ranging from Adam Smith to James Q. Wilson. These were the (older and newer) classics, whose conception of human nature as a moral and social being best fitted the data provided by modern science. Very importantly, these 'sociologists of virtue' were credited for showing 'that human beings in a free society will spontaneously manifest their natural moral sense' (Arnhart, 2005b, p. 35).

In essence, for Arnhart (Arnhart, 2005b, p. 17) this meant that the moral sense tradition rationalised, in philosophical and scientific terms, the essential element in both libertarian and traditionalist strands of conservative political thought, namely, 'the importance of spontaneous order'. As Arnhart explained, libertarians 'follow [Adam] Smith in stressing the spontaneous order of free markets', whereas traditionalists 'follow [Edmund] Burke in stressing the spontaneous order of customary practices. Both sides see spontaneous order as the only way in which social order can be achieved in a manner that is compatible with individual liberty' (Arnhart, 2005b, p. 17). Elaborating a few pages later on the nature of conservative – that is, his own ('fusionist') – political theory, Arnhart presented two of its foundational principles or demands. The first key demand was 'a political order in which government is limited to providing the legal conditions for people to govern themselves free from the arbitrary coercion of others' (Arnhart, 2005b, p. 35). The second key principle was a moral order in which 'the institutions of civil society – such as family life, economic exchange and religious activity – shape the character of citizens so that they have the moral capacity for governing themselves' (Arnhart, 2005b, p. 35). Although these two principles were the ideological backbone of Arnhart's fusionist conservatism, his philosophy was not complete without the moral sense theorists. Indeed, as Arnhart (Arnhart, 2005b, p. 35) put it, the moral order of civil society 'required' a 'sociology' that assumed 'that human beings in a free society will spontaneously manifest their natural moral sense'. This, of course, was, as we have seen, exactly what the moral sense/emotions tradition, or 'the conservative sociology of virtue' (Arnhart, 2005b, p. 35), as interpreted by Arnhart, assumed. We may thus conclude that in Arnhart's project of 'Darwinian conservatism', the best intellectual vindication of key conservative ideas was found in Darwinian theorists of the moral sense, among whom Arnhart counted – although he did not make it explicit in *Darwinian Conservatism* – Edward Westermarck. Because Arnhart's book was an explicit and powerful attack on the leftist ideology of strong central governance and planning, we may add that in Arnhart's case, as in Fukuyama's, Westermarck's work inspired a political theory that was leaning towards libertarianism.

Westermarck and Westermarckian conservatism

As already noted, some interpretations of Westermarck by American conservative intellectuals are controversial to say the least. It is necessary to assess them with reference to both Westermarck's own understanding of his work and recent Westermarck scholarship. I will focus on the three key themes critical for the 'Westermarckian' conservatives: (1) the notion of natural right, (2) the idea of the traditional family and (3) the libertarian-leaning interpretation of Westermarck's theory of incest.

First, let us consider Arnhart's construal of Westermarck as a natural right theorist. As noted earlier, the reading of Westermarck's moral theory as belonging to normative and objectivist ethics is highly contentious in the light of Westermarck's open advocacy of ethical relativism and his explicit dismissal of normative ethics as a science. Westermarck scholars generally agree that the Finnish anthropologist and philosopher was a moral relativist who attempted to show how and why human beings in general behaved *as if* objective moral facts did exist (Stroup, 1984, p. 576). As Timothy Stroup explains, Westermarck 'disagrees with the objectivist' (who accepts moral facts as true) and holds that 'We [human beings] try to ascribe moral objectivity but succeed only in objectivising our feelings'. The notion of 'objectivising' is thus 'crucial to understanding of Westermarck's view' (Stroup, 1984, p. 577; Westermarck, 1906, pp. 4–20). Accordingly, Westermarck himself described his approach as 'psychological ethics', that is, 'an inductive study of the actual conceptions of right and wrong', and contrasted it with normative ethics (Pipatti, 2011, p. 240; Stroup, 1984, esp. p. 576; Westermarck, [1896] 2003, p. 45).

Second, I wish to draw attention to the discrepancy between conservative uses of Westermarck's ideas on the family and recent studies of Westermarck in nineteenth-century intellectual and historical context. In light of historical studies that portray Westermarck as a reformist or even radical rather than traditionalist thinker, it is somewhat paradoxical that Westermarck's writings have been so enthusiastically cited by contemporary conservatives uneasy about the late-twentieth-century transformation of the family. The historian Elazar Barkan, for example, describes Westermarck as a writer who challenged the 'accepted [Victorian] etiquette' as regards marriage and sexuality, exemplifying 'the libertine scholarly position' (Barkan, 1995, pp. 63–4; see also Timosaari, 2010). While discussing Victorian attitudes to (primitive) sexuality and the belief in linear moral progress, Barkan notes that Westermarck's HHM put forward 'an egalitarian statement', which was 'also a subtle denigration of Victorian society'. That is to say, Westermarck challenged the widespread Victorian notion of the history of marriage as a progressive evolution from promiscuity to polyandry and polygamy, by arguing that 'the family (i.e., marriage) actually originated in the monogamous family' and that '"promiscuity" resulted from evolution'. These ideas, as illustrated in Westermarck's claim that 'in the lowest tribes chastity is more respected than in the higher ones', Barkan describes as 'scandalous' by contemporary standards. All in all, according to Barkan, Westermarck's notion of the origins of family should be understood as a 'radical' critique of received wisdom in the sense that the 'most celebrated institution' of Victorian society, the monogamous family, was 'identified with the most savage of unions' (Barkan, 1995, p. 64; Westermarck, 1921a, p. 159).

Thus, although Arnhart and others are right to assert that Westermarck saw marriage and the family as rooted in biological instincts, their attempts at recruiting Westermarck for the conservative side in the American 'war over the family' may disseminate an historically inaccurate image of Westermarck.[6]

Third, and lastly, it is in order to bring up Fukuyama's and Arnhart's interpretation of Westermarck's theory of the incest taboo. I will here leave open the question of whether one should call Westermarck a 'spontaneous order theorist' on the basis of his notion of moral norms as arising from emotions rather than from reason alone. However, it is evident that Westermarck did not draw similar libertarian-leaning, pro-civil society and antigovernment political conclusions from his moral theory as Fukuyama and Arnhart did. Of course, we will never know how Westermarck would have approached late-twentieth-century attempts to question the welfare state (however, see Salter, 2008).[7] From Westermarck's works we learn that at times he did underline individual responsibility, criticising oversized attempts at helping the needy, but it was not in the context of criticising the government for doing too much. For example, in *ODMI I* Westermarck (pp. 555–6) noted that in modern times 'the enlightened opinion' disagrees with 'unlimited [Christian] open-handedness' or 'indiscriminate almsgiving' due to its 'demoralizing influence' on the (poor) receiver.[8] But rooting for Seneca's and Cicero's prudent or reasonable forms of charity (Westermarck, 1906, pp. 554–5) is, of course, very different from being an antigovernment political writer or activist. One might ask, then, if there is any sense in regarding Westermarck himself as a 'conservative' writer or thinker. At the outset, it must be noted that 'conservatism' in itself is a historically evolving set of beliefs. Nevertheless, some students of conservatism do argue that one or two essential, unchanging core conservative ideas do exist. The following key notions have been suggested: belief in extra-human origins of the social order (such as God, history, biology or science) and an urge to render change safe (Freeden, 1996, pp. 332–47); hostility to emancipating the lower orders (Robin, 2011, esp. pp. 7–8); or, 'an *attitude* to social and political change that looks for support to the ideas, beliefs and habits of the past and puts more faith in the lessons of history than in the abstractions of political philosophy' (Allitt, 2009, p. 2). Because space does not allow me to go into detail on whether or not Westermarck's thought and writing fits some or all of these supposedly general conservative principles, I will restrict myself to a couple of short remarks most relevant to my theme, that is, Westermarck and late twentieth- and early twenty-first-century American conservatism.[9]

If one is to compare Westermarck with present-day American conservatives who root for moral absolutes, small government and the traditional family, it must be noted that Westermarck's regular endorsement of moral relativism and the lack of any systematic antigovernment rhetoric (despite his support for the idea of personal responsibility) do not speak for a deep ideological congruence between the Finn and his American followers. As to his views on family, it is to be noted that Westermarck, in stark contrast to many of his conservative American followers, explicitly argued for *liberalising* family law, for example relating to divorce.[10] And perhaps, most importantly, whereas Wilson, Arnhart, Fleming and Fukuyama were perceived by their contemporaries – and by themselves – as conservatives (on family issues), Westermarck was elected vice president of two rival feminist societies in Britain (Westermarck, 1929b, p. 98; for a particularly critical contemporary leftist interpretation, see, however, Calverton, 1931).

In conclusion, it appears that Westermarck's general political outlook contrasts to a considerable extent with the ideology of his later American followers, namely the post–Second World War 'conservatism' defined by the historian George H. Nash (2006, p. xix) as 'resistance to certain forces perceived to be leftist, revolutionary, and profoundly subversive of what conservatives at the time deemed worth cherishing, defending, and perhaps dying for'. Nevertheless, as we have just seen, Westermarck's scientific works are, and have always been, open to various types of political use, at times different from what Westermarck probably could have imagined. Such openness to interpretation and exploitation, of course, holds true of scientific writing in general.

Notes

1 First known for its social scientific analyses of the (in)efficiency of the big welfare state initiatives in the late 1960s and early 1970s, the neoconservative movement became increasingly associated with a moral approach to America's social problems.

2 I wish to thank Larry Arnhart for allowing me to quote from his lecture.

3 Fukuyama is best known for his 1992 bestseller *The End of History and the Last Man*. He long regarded himself a neoconservative, but distanced himself from the group after the 2003 Iraq invasion.

4 Fukuyama (1999, p. 16) defined 'social capital' as 'a set of informal values or norms shared among members of a group that permits cooperation among them'.

5 Arnhart does not always clearly separate the concepts of 'the moral sense' and 'moral emotions.' Westermarck in his mature moral theory, however, was much more consistent in his use of the concept of moral emotions. See Piipatti, Chapter 10.

6 Not all historians agree that Westermarck's books on family had a reformist bent, however. To take one example, Greta Jones (1982, p. 125) suggests that a feel of 'comforting reassurance' existed in Westermarck's HHM, because it seemed to guarantee that 'socially desirable behaviour would continue in a world of shifting values'. Lyons and Lyons (2004, p. 113) suggest that it is easy to find both modern relativist views and Victorian ones in Westermarck's works.

7 Salter (2008, p. 28) argues that Westermarck's 'style of analysis can be applied to exploring the links, if any, between charitableness and the welfare state as an institution'.

8 I thank Niina Timosaari for bringing into my attention Westermarck's ideas on charity.

9 As Olli Lagerspetz (2013, p. 919) and many others have noted, Westermarck was truly active in public matters on at least two occasions: In the period of 'Russification' of the Grand Duchy of Finland at the turn of the twentieth century (before Finland's declaration of independence in 1917), Westermarck used his international contacts to get scientists' support for the Finnish case. Later, in 1921, when the League of Nations discussed the question whether Åland (Ahvenanmaa) should belong to Sweden or Finland, Westermarck, as part of the Finnish delegation, argued for the islands being retained by Finland. In Britain, Westermarck had many contacts with liberal and leftist intellectuals and activists (e.g., the Fabian Society, the Bloomsbury group, and liberal donors), as Lagerspetz and Suolinna (2014, esp. chapter 2) note.

10 Niina Timosaari discusses this theme at length in her forthcoming doctoral dissertation 'Sukupuolta ja seksuaalisuutta koskevien lakien muutostarpeet Edvard Westermarckin (1862–1939) tieteellisessä tuotannossa'. She has generously allowed me to read a draft of a particularly relevant chapter.

Bibliography

Abbreviations of Westermarck's works

ER: 1932. *Ethical relativity.*
HHM: 1891. *The history of human marriage.*
HHM I–III: 1921 *The history of human marriage*, 5th ed., Vols. 1–3.
MML: 1929. *Memories of my life.*
ODMI I–II: 1906–1908. *The origin and development of the moral ideas*, Vols. 1–2.
RBM I–II: 1926. *Ritual and belief in Morocco*, Vols. 1–2.
SHM: 1926. *A short history of marriage.*

Manuscript material

Åbo Akademi university library, manuscript and picture unit collections

ÅAB: EW. Edvard Westermarcks brevsamling [Letters to Edward Westermarck].
ÅAB: HW. Helena Westermarcks brevsamling [Letters to Helena Westermarck].
ÅAB: RL. Rolf Lagerborgs brevsamling [Letters to Rolf Lagerborg].
ÅAB: KP. Knut Pippings brevsamling [Letters to Knut Pipping].

Ihanus, J., Lahtinen, T., & Neuman, Y. (eds.), 2011. *The Edward Westermarck online collection in Filosofia.fi*. Available at: http://filosofia.fi/se/arkiv.

Westermarck, E., 1897. *Till Consistorium, Anhållan om Hermann Rosenbergs stipendium* [To the University Senate, application for Hermann Rosenberg grant]. Westermarck Collection Box 42a, Academica, affärspapper. Åbo: Åbo Akademi University Library Manuscript and Picture Unit collections.

Westermarck, E., 1899–1923. *The 'record note book': Aforismer från lediga stunder 1899–1923*. Åbo: Åbo Akademi University Library Manuscript and Picture Unit collections.

Westermarck, E., 1913. *Föreläsningar 1913, Hume* [Lectures 1913]. Westermarck Collection Box 78, Föreläsningar [Lectures]. Åbo: Åbo Akademi University Library Manuscript and Picture Unit collections.

Westermarck, E., 1914. *Föreläsning i filosofins historia: vårterminen 1914, Adam Smith* [Lecture in the history of philosophy, spring term 1914]. Westermarck's manuscript collection Box 78, Föreläsningar [Lectures]. Åbo: Åbo Akademi University Library Manuscript and Picture Unit collections.

The National Library of Finland, manuscript collection

NLF: KT Knut Tallqvists brevsamling [Letters to Knut Tallqvist].

The National Archive of Finland

Hagelin, S., 1897. *Psykologi: Anteckningar till dr. E. A. Westermarcks föreläsningar*. [Psychology. Notes according to Dr. E. A. Westermarck's lectures]. Helsinki: The National Archive of Finland.

Other

Arnhart, L., 1998a. Westermarck's ethics as Darwinian natural right. Plenary lecture, International Westermarck Symposium. Helsinki, Finland, 19–22 November (manuscript).

BMC: Bronislaw Malinowski Collection. General correspondence. British library of political and economic science. London: School of Economics and Political Science.

BMP: Bronislaw Malinowski Papers. New Haven, Connecticut: Yale University Library.

Concise catalogue of Freud's library (s.d.). London: Freud museum

MA: Macmillan Archive. London: British Library.

Published sources

Works by Edward Westermarck

Westermarck, E., [1890] 1996. Darwins hypotes beträffande den moraliska känslans uppkomst ['Darwin's hypothesis concerning the origin of the moral sense']. In: Manninen, J., & Niiniluoto, I., eds., *Ajatuksen laboratorio: Suomen Filosofisen Yhdistyksen pöytäkirjat 1873–1925*. Helsinki: Philosophical Society of Finland, pp. 160–161.

HHM: Westermarck, E., 1891a. *The history of human marriage*. London: Macmillan and Co.

Westermarck, E., 1891b. Selektionsteorin och dess betydelse för vetenskaperna om det fysiska, psykista och sociala livet ['The theory of selection and its significance to the physical, mental and social life']. *Album utgifvet af Nyländningar* (10). Helsingfors: Waseniuska Bokhandeln, pp. 218–240.

Westermarck, E., [1892] 1996. Tahdon vapaus ja sen siveellinen merkitys ['Freedom of will and its ethical relevance']. In: Manninen, J., & Niiniluoto, I., eds., *Ajatuksen laboratorio: Suomen Filosofisen Yhdistyksen pöytäkirjat 1873–1925*. Helsinki: Philosophical Society of Finland, pp. 194–197.

Westermarck, E., 1894. *Anteckningar i samhällslära enligt Dr. Edv. Westermarcks föreläsningar.* Helsingfors: Akademiska Bokhandeln.

Westermarck, E., [1895] 1996. Den moraliska känslan ['The moral feeling']. In: Manninen, J., & Niiniluoto, I., eds., *Ajatuksen laboratorio: Suomen Filosofisen Yhdistyksen pöytäkirjat 1873–1925*. Helsinki: Philosophical Society of Finland, pp. 237–244.

Westermarck, E., 1896. Sinnesrörelserna och deras kroppsliga uttryck. *Finsk Tidskrift*, 40(5), pp. 338–353.

Westermarck, E., [1897] 1996. Predikatet i moraliska omdömen ['The Predicate in moral judgements']. In: Manninen, J., & Niiniluoto, I., eds., *Ajatuksen laboratorio: Suomen Filosofisen Yhdistyksen pöytäkirjat 1873–1925*. Helsinki: Philosophical Society of Finland, pp. 298–303.

Westermarck, E., 1898. The essence of revenge. *Mind*, 7(27), pp. 289–310.

Westermarck, E., 1900. Remarks on the predicates of moral judgments. *Mind*, 9(34), pp. 184–204.

Westermarck, E., 1905. Religion och magi. *Euterpe*, 1905(3–4), pp. 21–29.

ODMI I: Westermarck, E., 1906a. *The origin and development of the moral ideas*. Vol. 1. London: Macmillan & Co.

Westermarck, E., 1906b. The influence of magic on social relationships: Read before a meeting of the sociological society, March 22nd, 1905. *Sociological Papers*, II, pp. 143–174.

Westermarck, E., 1907. *Religion och magi: Studentföreningen Verdandis småskrifter 149*. Stockholm: Albert Bonniers förlag.

ODMI II: Westermarck, E., 1908. *The origin and development of the moral ideas*. Vol. 2. London: Macmillan & Co.

Westermarck, E., 1909. *Sexualfragen: Übersetzt von Leopold Katscher*. Leipzig: Verlag von Dr. Werner Klinkhardt.

Westermarck, E., 1913. [*Review of Malinowski, The Family among the Australian Aborigines*]. *Folk-Lore*, 24(3), pp. 406–408.

Westermarck, E., 1914. *Marriage ceremonies in Morocco*. London: Macmillan.

Westermarck, E., 1916. *The Moorish conception of holiness (baraka)*. Öfversigt af Finska Vetenskaps-Societetens förhandlingar Bd. LVIII, 1915–1916, Afd. B. No 1, Helsingfors: Societas scientiarum fennica.

Westermarck, E., 1918. *Sex år i Marocko*. Helsingfors: Schildts.

HHM I, 1921: Westermarck, E., 1921a. *The history of human marriage*. Vol. 1. 5th ed. London: Macmillan and Co.

HHM II, 1921: Westermarck, E., 1921b. *The history of human marriage*. Vol. 2. 5th ed. London: Macmillan and Co.

HHM III, 1921: Westermarck, E., 1921c. *The history of human marriage*. Vol. 3. 5th ed. London: Macmillan and Co.

Westermarck, E., 1921d. *The origin of sexual modesty: The British society for the study of sex psychology*. Vol. 8. London: J.E. Francis.

Westermarck, E., 1925. *Studieresor i ungdomen. K.D.F. [Festskrift utgiven av Nylänningar]*, pp. 26–30. Helsingfors: Nyländska avdelningen.

RBM I, RBM II: Westermarck, E., 1926a. *Ritual and belief in Morocco*. Vols. 1–2. London: Macmillan and Co.

SHM: Westermarck, E., 1926b. *A short history of marriage*. London: Macmillan and Co.

Westermarck, E., 1928. Beliefs relating to sexual matters in Morocco. In Marcuse, M., ed., *Verhandlungen des I. internationalen Kongresses für Sexualforschung*, Vol. 5. Band, pp. 163–169. Berlin: A. Marcus & E. Weber.

Westermarck, E., 1928–9. *L'origine et le développement des idées morales*. Traduction de Robert Godet. Paris: Payot.

Westermarck, E., 1929a. *Marriage*. London: Benn's Sixpenny Library.

MML: Westermarck, E., 1929b. *Memories of my life*. Translated by Anna Barwell. London: George Allen & Unwin.

Westermarck, E., 1930. *Wit and wisdom in Morocco: A study of native proverbs. By Edward Westermarck, with the assistance of Shereef 'Abd-Es-Salam El-Baqqali*. London: Routledge.

ER: Westermarck, E., 1932a. *Ethical relativity*. London: Kegan Paul.

Westermarck, E., 1932b. *Early beliefs and their social influence*. London: Macmillan and Co.

Westermarck, E., 1932c. The study of popular sayings. In: Dawson, W.R., ed., *The Frazer Lectures 1922–1932*. London: Macmillan, pp. 190–211.

Westermarck, E., 1933. *Pagan survivals in Mohammedan civilization*. London: Macmillan.

Westermarck, E., 1934a. *Freuds teori om Oedipuskomplexen i sociologisk belysning*. Vetenskap och bildning. Vol. 45. Stockholm: Albert Bonniers Förlag.

Westermarck, E., 1934b. *Three essays on sex and marriage*. London: Macmillan and Co.

Westermarck, E., 1936. Methods in social anthropology. *Journal of the Royal Anthropological Institute*, 66, pp. 223–248.

Westermarck, E., 1939. *Christianity and morals*. London: Kegan Paul, Trench, Trubner & Co.

Westermarck, E., 2003. Normative and psychological ethics. In: Haaparanta, L., & Niiniluoto, I., eds., *Analytic philosophy in Finland*. Amsterdam: Rodopi, pp. 43–47.

Westermarck, E., 2014a. Implications of the theory of selection (1889). Edited and translated by Antfolk, J. In: Shankland, D., ed., *Westermarck*. Occasional Paper No. 44 of the Royal Anthropological Institute. Canon Pyon: Sean Kingston Publishing, pp. 147–161.

Westermarck, E., 2014b. Survival in ritual: Delivered to the International Congress of Ethnological and Anthropological Sciences (1934), Shankland, D. In: Shankland, D., ed., *Westermarck*. Occasional Paper No. 44 of the Royal Anthropological Institute. Canon Pyon: Sean Kingston Publishing, 2014, pp. 169–177.

Works by others

Ackerman, J. M., Kenrick, D. T., & Schaller, M., 2007. Is friendship akin to kinship? *Evolution and Human Behavior*, 28, pp. 365–374.

Adams, M. S., & Neel, J. V., 1967. Children of incest. *Pediatrics*, 40(1), pp. 55–62.

Allardt, E., 2000. Edward Westermarck: A sociologist relating nature and culture. *Acta Sociologica*, 43, pp. 299–306.

Allitt, P., 2009. *The Conservatives: Ideas and personalities throughout American history*. New Haven: Yale University Press.

Alper, J. S., & Lange, R. V., 1981. Lumsden-Wilson theory of gene-culture coevolution. *Proceedings of the National Academy of Science*, 78(6), pp. 3976–3979.

Anscombe, G. E. M., 1981a. Modern moral philosophy. In: Anscombe, G. E. M., ed., *The collected philosophical papers of G.E.M. Anscombe. Volume III: Ethics, religion and politics*. Oxford: Basil Blackwell, pp. 26–42.

Anscombe, G. E. M., 1981b. On brute facts. In: Anscombe, G. E. M., ed., *The collected philosophical papers of G.E.M. anscombe. Volume III: Ethics, religion and politics*. Oxford: Basil Blackwell, pp. 22–25.

Antfolk, J., Karlsson, M., Bäckström, A., & Santtila, P., 2012. Disgust elicited by third-party incest: The roles of biological relatedness, co-residence, and family relationship. *Evolution and Human Behavior*, 33(3), pp. 217–223.

Antfolk, J., Lieberman, D., Albrecht, A., & Santtila, P., 2014. The self-regulation effect of fertility status on inbreeding aversion: When fertile, disgust increases more in response to descriptions of one's own than of others' inbreeding. *Evolutionary Psychology*, 12(3), pp. 621–631.

Antfolk, J., Lieberman, D., & Santtila, P., 2012. Fitness costs predict inbreeding aversion irrespective of self-involvement: Support for hypotheses derived from evolutionary theory. *PLoS ONE*, 7(11), pp. 1–7.

Antfolk, J., Lindqvist, H., Albrecht, A., & Santtila, P., 2014. Self-reported availability of kinship cues during childhood is associated with kin-directed behavior to parents in adulthood. *Evolutionary Psychology*, 12(1), pp. 148–166.

Antfolk, J., Santtila, P., Harju, C., Albrecht, A., Mokros, A, & Lieberman, D., submitted. The human mind computes the inclusive fitness costs in regulating inbreeding.

Aoki, K., 2004. Avoidance and prohibition of brother–sister sex in humans. *Population Ecology*, 47(1), pp. 13–19.

Arnhart, L., 1998b. *Darwinian natural right: The biological ethics of human nature*. Albany: State University of New York Press.

Arnhart, L., 2001. Thomistic natural law as Darwinian natural right. In: Paul, E. F. et al., eds., *Natural law and modern moral philosophy*. Cambridge: Cambridge University Press, pp. 1–33.

Arnhart, L., 2003. Darwinian conservatism as the new natural law. *The Good Society*, 12(3), pp. 14–19.

Arnhart, L., 2005a. The incest taboo as Darwinian natural right. In: Wolf, A. P., & Durham, W. H., eds., *Inbreeding, incest, and the incest taboo: The state of knowledge at the turn of the century*. Stanford: Stanford University Press, pp. 190–217.

Arnhart, L., 2005b. *Darwinian conservatism*. Charlottesville, VA: Imprint Academic.

Arnhart, L., 2006. The new fusionism. *Darwinian conservatism blog*. [blog] 8 December. Available at: http://darwinianconservatism.blogspot.fi/2006/12/new-fusionism.html [accessed 24 May 2015].

Arnhart, L., 2007. John West's response, Parts 1–2. *Darwinian conservatism blog*, [blog] 8 January. Available at: http://darwinianconservatism.blogspot.fi/2007/01/john-wests-response-parts-1–2.html [accessed 25 May 2015].

Arnhart, L., 2011. Darwinian marriage (2): A response to Robert George. *Darwinian conservatism blog*, [blog] 28 January. Available at: http://darwinianconservatism.blogspot.fi/2011/01/darwinian-marriage-2-response-to-robert.html [accessed 24 May 2015].

Ayer, A. J., [1936] 1971. *Language, truth and logic*. Harmondsworth: Penguin Books.

Backström, J., 2007. *The fear of openness: An essay on friendship and the roots of morality*. Åbo: Åbo Akademi University Press.

Backström, J., & Nykänen, H., 2015. Collectivity, evil and the dynamics of moral value. *Journal of Evaluation in Clinical Practice*. DOI: 10.1111/jep.12311.

Badcock, C. R., 1980. *The psychoanalysis of culture*. Oxford: Basil Blackwell.

Badcock, C. R., 1994. *PsychoDarwinism: The new synthesis of Darwin & Freud*. London: HarperCollins.

Bagnoli, C., 2011. Emotions and the categorical authority of moral reason. In: Bagnoli, C., ed., *Morality and the emotions*. Oxford: Oxford University Press, pp. 62–81.

Barkan, E., 1995. Victorian promiscuity: Greek ethics and primitive exemplars. In: Barkan, E., & Bush, R., eds., *Prehistories of the future: The Primitivist project and the culture of the modern*. Stanford: Stanford University Press, pp. 56–92.

Bateson, P., 2005. Inbreeding avoidance and incest taboos. In: Wolf, A. P., & Durham, W. H., eds., *Inbreeding, incest, and the incest taboo: The state of knowledge at the turn of the century*. Stanford, CA: Stanford University Press, pp. 24–37.

Bennett, M. R., & Hacker, P. M. S., 2003. *Philosophical foundations of neuroscience*. Oxford: Blackwell Publishing.

Bevc, I., & Silverman, I., 1993. Early proximity and intimacy between siblings and incestuous behavior: A test of the Westermarck hypothesis. *Ethology and Sociobiology*, 14(3), pp. 171–181.

Bevc, I., & Silverman, I., 2000. Early separation and sibling incest: A test of the revised Westermarck theory. *Evolution and Human Behavior*, 21(3), pp. 151–161.

Bittles, A. H., 1983. The intensity of human inbreeding depression. *Behavioral and Brain Sciences*, 6(1), pp. 103–104.

Blankenhorn, D., 2009. *The future of marriage*. New York: Encounter Books.

Bloch, M., 1998. Internal and external memory: Different ways of being in history. In: Siikala, J. et al., eds., *Developing Anthropological Ideas: The Westermarck Memorial Lectures 1983–97*. Transactions of the Finnish Anthropological Society No. 41, Helsinki: The Finnish Anthropological Society, pp. 143–166.

Blum, L., 2011. Empathy and empirical psychology: A critique of Shaun Nichols's neo-sentimentalism. In: Bagnoli, C., ed., *Morality and the emotions*. Oxford: Oxford University Press, pp. 170–196.

Bourguignon, E., 1979. *Psychological anthropology: An introduction to human nature and cultural differences*. New York: Holt, Rinehart and Winston.

Bourquia, R., & Al Harras, M., eds., 1993. *Westermarck et la société marocaine*. Rabat: Université Mohammed V, Publications de la Faculté des Lettres et des Sciences Humaines – Rabat. Série Colloques et Séminaires no 27.

Bowman, M., 2006. Wilson, James Q. In: Frohnen, B., Beer, J. and Nelson J.O., eds., *American conservatism: An encyclopedia*. Wilmington: ISI Books, p. 928.

Broackes, J., ed., 2011. *Iris Murdoch: Philosopher*. Oxford: Oxford University Press.

Broadie, A., 2009. *A history of Scottish philosophy*. Edinburgh: Edinburgh University Press.

Brown, K., 1982. The 'curse' of Westermarck. In: Stroup, T., ed., *Edward Westermarck: Essays on his life and works*. Acta Philosophica Fennica. Vol. 34. pp. 219–259.

Burton, R., 1973. Folk theory and the incest taboo. *Ethos*, 1, pp. 504–516.

Calverton, V. F., 1931. The compulsive basis of social thought: As illustrated by the varying doctrines as to the origins of marriage and the family. *The American Journal of Sociology*, 36(5), pp. 689–720.

Cavalli-Sforza, L. L., & Feldman, M. W., 1981. *Cultural transmission and evolution*. Princeton, N. J.: Princeton University Press.

Cockburn, D., 1995. Responsibility and necessity. *Philosophy*, 70(273), pp. 409–427.

Cockburn, D., 2001. *An introduction to the philosophy of mind*. Basingstoke: Palgrave Macmillan.

Cockburn, D., 2009. Emotion, expression and conversation. In: Gustafsson, Y., Kronqvist, C., & McEachrane, M., eds., *Emotions and understanding: Wittgensteinian perspectives*. Basingstoke: Palgrave, pp. 126–144.

Comte, A., 1858. *The catechism of positive religion*. Translated by Congreve, R., London: Chapman.

Cook, J. W., 1999. *Morality and cultural differences*. Oxford: Oxford University Press.

Cosmides, L., & Tooby, J., 1992. Cognitive adaptations for social exchange. In: Barkow, J. H., Cosmides, L., & Tooby, J., eds., *The adapted mind: Evolutionary psychology and the generation of culture*. New York: Oxford University Press, pp. 163–228.

Cosmides, L., & Tooby, J., 2000. Evolutionary psychology and the emotions. In: Lewis, M., & Haviland-Jones, J. M., eds., *Handbook of emotions*. 2nd Edition. New York: Guilford, pp. 91–115.

Cosmides, L., & Tooby, J., 2005. Social exchange: The evolutionary design of a neurocognitive system. In: Gazzaniga, M., ed., *The new cognitive neurosciences, III*. Cambridge, MA: MIT Press, pp. 1295–1308.

Cosmides, L., & Tooby, J., 2008. Can a general deontic logic capture the facts of human moral reasoning? How the mind interprets social exchange rules and detects cheaters. In: Sinnott-Armstrong, W., ed., *Moral psychology, Vol 1, The evolution of morality: Adaptations and innateness*. Cambridge: The MIT Press, pp. 53–119.

Crews, F., 1996. The verdict on Freud. *Psychological Science*, 7(2), pp. 63–68.

Darwin, C., [1859,1871] 1872. *The origin of species and the descent of man*. 6th edition. New York: The Modern Library.

Darwin, C., [1871] 2010. *The descent of man*. Abridged by Ghiselin, M., New York: Dover Publications.

Darwin, C., 1981. *The descent of man, and selection in relation to sex*. Princeton, NJ: Princeton University Press.

Dawkins, R., 2006. *The selfish gene*. 30th anniversary edition. Oxford: Oxford University Press.

De Caro, M., & Macarthur, D., 2010. Science, naturalism and the problem of normativity. In: De Caro, M., & Macarthur, D., eds., *Naturalism and normativity*. New York: Columbia University Press, pp. 1–19.

Dennett, D. C., 2004. *Freedom evolves*. London: Penguin Books.

Descombes, V., 2001. *The mind's provisions: A critique of cognitivism*. Princeton: Princeton University Press.

De Waal, F., 1996. *Good natured: The origins of right and wrong in primates and other animals*. Cambridge, MA: Harvard University Press.

De Waal, F., 2006. Morally evolved: Primate social instincts, human morality and the rise and fall of 'veneer theory'. In: Macedo, S., & Ober, J., eds., *Primates and philosophers: How morality evolved*. Princeton: Princeton University Press, pp. 1–80.

De Waal, F., 2008. Putting the altruism back into altruism: The evolution of empathy. *Annual Review of Psychology*, 59, pp. 279–300.

De Waal, F., 2009. *The age of empathy: Nature's lessons for a kinder society*. New York: Harmony Books.

Dixon, T., 2003. *From passions to emotions: The creation of a secular psychological category*. Cambridge: Cambridge University Press.

Du Bois, C., 1937. Anthropological perspectives on psychoanalysis. *Psychoanalytic Review*, 24(3), pp. 246–263.

Durham, W. H., 1978. Toward a coevolutionary theory of human biology and culture. In: Caplan, A., ed., *The sociobiology debate*. New York: Harper & Row, pp. 428–448.

Durham, W. H., 1990. *Coevolution*. Stanford, CA: Stanford University Press.

Durham, W. H., 1991. *Coevolution: Genes, Culture, and Human Diversity*. Stanford: Stanford University Press.

Durham, W. H., 2005. Assessing the gaps in Westermarck's theory. In: Wolf, A. P., & Durham, W. H., eds., *Inbreeding, incest, and the incest taboo: The state of knowledge at the turn of the century*. Stanford, CA: Stanford University Press, pp. 121–138.

Durkheim, É., 1899. De la définition des phénomènes religieux. *L'Année sociologique*. Vol. II (1897–1898). Paris: Félix Alcan, pp. 1–28.

Durkheim, É., 1907. [Review of] Edward Westermarck: The origin and development of the moral ideas. *L'Année sociologique*, 10, pp. 383–395.

Durkheim, É., [1912] 1968. *The elementary forms of the religious life*. London: Allen & Unwin.

Eder, M. D., 1935. [Review of Westermarck 1932b]. *International Journal of Psycho-Analysis*, 16, pp. 115–116.

El-Mansour, M., 1999. The sanctuary (hurm) in precolonial Morocco. In: Bourquia, R., & Miller, S. G., eds., *In the shadow of the Sultan: Culture, power, and politics in Morocco*. Cambridge, MA: Center for the Middle Eastern Studies of Harvard University, pp. 49–73.

Ember, M., 1975. On the origin and extension of the incest taboo. *Behavior Science Research*, 10, pp. 249–281.

Erickson, M., 1989. Incest avoidance and familial bonding. *Journal of Anthropological Research*, 45, pp. 267–291.

Erickson, M., 2005. Evolutionary thought and the current clinical understanding of incest. In: Wolf, A. P., & Durham, W. H., eds., *Inbreeding, incest, and the incest taboo: The state of knowledge at the turn of the century*. Stanford, CA: Stanford University Press, pp. 161–189.

Fessler, D. M. T., 2007. Neglected natural experiments germane to the Westermarck hypothesis. *Human Nature*, 18(4), pp. 355–364.

Fessler, D. M. T., & Navarrete, C. D., 2004. Third-party attitudes toward sibling incest: Evidence for Westermarck's hypotheses. *Evolution and Human Behavior*, 25(5), pp. 277–294.

Firth, R., 1981. Bronislaw Malinowski. In: Silverman, S., ed., *Totems and teachers: Perspectives on the history of anthropology*. New York: Columbia University Press, pp. 100–139.

Flack, J. C., & de Waal, F. B. M., 2000a. 'Any animal whatever': Darwinian building blocks of morality in monkeys and apes. *Journal of Consciousness Studies*, 7, pp. 1–29.

Flack, J. C., & de Waal, F. B. M., 2000b. Being nice is not a building block of morality (Response to commentary discussion). *Journal of Consciousness Studies*, 7, pp. 67–77.

Fleming, T., [n.d.]. The Rockford Institute. *Chronicles: A Magazine of American Culture* [e-journal]. Available at: http://www.chroniclesmagazine.org/about/the-rockford-institute/ [accessed 7 October 2014].

Fleming, T., 1988. *The politics of human nature*. New Brunswick: Transaction Publishers.

Fletcher, D., 1982. On the contribution of Edward Westermarck. The process of institutionalization: A general theory. In: Stroup, T., ed., *Edward Westermarck: Essays on his Life and Works*. Helsinki: Acta Philosophica Fennica. Vol. 34, pp. 194–217.

Flügel, J. C., [1921] 1929. *The psychoanalytic study of the family*. London: Hogarth Press.

Flügel, J. C., 1935. [Review of Westermarck 1934a]. *International Journal of Psycho-Analysis*, 16, pp. 378–383.

Fortes, M., [1957] 1968. Malinowski and the study of kinship. In: R.Firth, R., ed., *Man and culture*. London: Routledge & Kegan Paul, pp. 157–188.

Fowler, H. W., & Fowler, F. G. [1911,1950] 1956. *Concise Oxford dictionary of current English*. Oxford: Clarendon Press.

Fowler, R. B., 1999. *Enduring liberalism: American political thought since the 1960s*. Lawrence: University Press of Kansas.

Francis, S., 2002. The Paleo persuasion. *The american conservative* [e-journal], 16 December. Available at: http://www.theamericanconservative.com/articles/the-paleo-persuasion/ [accessed 5 October 2014].

Frazer, J., 1910. *Totemism and exogamy*. London: Macmillan.

Frazer, J. G., 1911. *The magic art and the evolution of kings, Vol. I*. Vol. I (Part I.1) of *The golden bough. A study in magic and religion*. 3rd edition. London: Macmillan.

Frazer, J. G., 1913. *Balder the beautiful, Vol. II*. Vol. XI (Part VII.2) of *The golden bough. A study in magic and religion*. 3rd edition. London: Macmillan.

Freeden, M., 1996. *Ideologies and political theory: A conceptual approach*. Oxford: Oxford University Press.

Frege, G., 1893. *Grundgesetze der Arithmetik: Begriffsschriftlich abgeleitet, Vol. 1*. Jena: Pohle.

Freud, S., [1909] 1976. Der Familienroman der Neurotiker. *Gesammelte Werke*. Eds. Freud, A., Bibring, E.,Hoffer, W., Kris, E. & Isakower, O., Vol. VII. Frankfurt am Main: S. Fischer, pp. 227–231.

Freud, S., [1912–1913] 1946. *Totem and taboo*. New York: Vintage Books.
Freud, S., [1912–1913] 1978. *Totem und Tabu. Über einige Übereinstimmungen im Seelenleben der Wilden und der Neurotiker. Gesammelte Werke* IX. Frankfurt am Main: S. Fischer.
Freud, S., [1938] 1978. Abriss der psychoanalyse. *Gesammelte werke* XVII. Frankfurt am Main: S. Fischer, pp. 67–138.
Freud, S., [1939] 1978. *Der Mann Moses und die monotheistische Religion. Gesammelte Werke* XVI. Frankfurt am Main: S. Fischer, pp. 103–246.
Fukuyama, F., 1999. *The great disruption: Human nature and the reconstitution of social order*. London: Profile Books.
Fukuyama, F., 2006. *America at the crossroads: Democracy, power, and the neoconservative legacy*. New Haven: Yale University Press.
Gallie, W. B., 1956. Essentially contested concepts. In: Black, M., ed., 1962. *The importance of language*. Englewood Cliffs, NJ: Prentice-Hall, pp. 121–146.
Gaskin, R., 2006. *Experience and the world's own language (A critique of John McDowell's empiricism)*. Oxford: Oxford University Press.
Gellner, E., 1969. *Saints of the Atlas*. Chicago: The University of Chicago Press.
Gilder, G., 1973. *Sexual suicide*. New York: Quadrangle.
Godelier, M., 1989. Incest taboo and the evolution of society. In: Grafen, A. ed., *Evolution and its influence: The Herbert Spencer lectures 1986*. Oxford: Clarendon Press, pp. 63–92.
Gould, S. J., 1977. *Ontogeny and phylogeny*. Cambridge: Harvard University Press.
Gouldner, A., 1971. *The coming crisis of Western sociology*. London: Heinemann.
Gupta, A., & Ferguson, J., 1992. Beyond 'culture': Space, identity and the politics of difference. *Cultural Anthropology*, 7(1), pp. 6–23.
Gustafsson, Y., 2012. Three perspectives on altruism. In: Forsberg, N., Burley, M., & Hämäläinen, N., eds., *Language, ethics and animal life: Wittgenstein and beyond*. New York: Bloomsbury, pp. 82–110.
Gustafsson, Y., 2014. *Interpersonal understanding and theory of mind*. Åbo: Åbo Akademi University Press.
Haidt, J., 2001. The emotional dog and its rational tail: A social intuitionist approach to moral judgment. *Psychological Review*, 108(4), pp. 814–834.
Haidt, J., 2012. *The righteous mind: Why good people are divided by politics and religion*. New York: Pantheon Books.
Haig, D., 1999. Asymmetric relations: Internal conflicts and the horror of incest. *Evolution and Human Behavior*, 20(2), pp. 83–98.
Hamilton, W. D., 1964. The genetical evolution of social behaviour: I. *Journal of Theoretical Biology*, 7, pp. 1–16.
Harris, W., [1921] 1983. *Morocco that was*. London: Eland Books.
Hartung, J., 1985. [Review of Shepher, 1983]. *American Journal of Physical Anthropology*, 67(2), pp. 169–171.
Hempel, C. G., 1965. *Aspects of scientific explanation and other essays in the philosophy of science*. New York: The Free Press.
Herle, A., & Rouse, S., eds., 1998. *Cambridge and the Torres Strait: Centenary essays on the 1898 anthropological expedition*. Cambridge: Cambridge University Press.
Hertzberg, L., 2011. Very general facts of nature. In: Kuusela, O., & McGinn, M., eds., *The Oxford handbook of Wittgenstein*. Oxford: Oxford University Press, pp. 351–374.
Hertzberg, L., 2013. On knowing right from wrong. In: Gustafsson, Y., Kronqvist, C., & Nykänen, H., eds., *Ethics and the philosophy of culture: Wittgensteinian approaches*. Cambridge: Cambridge Academic Publishers, pp. 160–170.

Hirn, Y., 1947. Edward Westermarck and his English friends. *Transactions of the Westermarck Society*, 1(2), pp. 39–51.

Hobbes, T., [1651] 1985. *Leviathan*. London: Penguin.

Hobson, P., 2003. *The cradle of thought*. Oxford: Oxford University Press.

Hollingsworth v. Perry, 2013. Brief of National Association of Evangelicals et al. as amici curiae in support of petitioners, Hollingsworth v. Perry, 133 S. Ct. 2652 (January 29, 2013) (No. 12–144). Available at: http://www.americanbar.org/content/dam/aba/publications/supreme_court_preview/briefs-v2/12–144_pet_amcu_nae-etal.authcheckdam.pdf [accessed 23 June 2015].

Hook, R. H., 1979. Phantasy and symbol: A psychoanalytic point of view. In: Hook, R. H., ed., *Fantasy and symbol. Studies in anthropological interpretation*. London: Academic Press, pp. 267–291.

Hume, D., 1739. *A treatise of human nature*. Vol. I-III. Oxford: Clarendon Press.

Hume, D., 1751. *An enquiry concerning the principles of morals*. Oxford: Clarendon Press.

Husserl, E., 1900. Logische Untersuchungen. Erster Band: Prolegomena zur reinen Logik. In: Ed. Holenstein, E., 1975. The Hague: Nijhoff.

Hutchinson, P., 2009. Emotion-philosophy-science. In: Gustafsson, Y., Kronqvist, C., & McEachrane, M., eds., *Emotions and understanding: Wittgensteinian perspectives*. Basingstoke: Palgrave, pp. 60–80.

Ihanus, J., 1998. Westermarck, sexology and the history of human marriage. In: Brinkhuis, F., & Talmor, S., eds., *Memory, history and critique: European identity at the millennium. Proceedings of the 5th Conference of the International Society of the Study of European Ideas, Utrecht, August 19–24, 1996* (CDROM). Cambridge, MA: ISSEI & The MIT Press.

Ihanus, J., 1999. *Multiple origins: Edward Westermarck in search of mankind*. Frankfurt am Main: Peter Lang.

James, W., 1884. What is an emotion? *Mind*, 9(34), pp. 188–205.

James, W., 1890. *Principles of psychology*. New York: Henry Holt.

Jones, E., 1922. [Review of HHM I-III]. *International Journal of Psycho-Analysis*, 3, pp. 249–252.

Jones, E., 1924. Psychoanalysis and anthropology. *Journal of the Royal Anthropological Institute*, 54, pp. 47–67.

Jones, E., 1925. Mother right and the sexual ignorance of savages. *International Journal of Psycho-Analysis*, 6, pp. 109–130.

Jones, E., 1957. *The life and work of Sigmund Freud*. Vol. 3: *The last phase, 1919–1939*. London: Hogarth Press.

Jones, G., 1982. *Social Darwinism and English thought: The interaction between biological and social theory*. Brighton: The Harvester Press.

Joyce, R., 2008. What neuroscience can (and cannot) contribute to metaethics. In: Sinnott-Armstrong, W., ed., *Moral psychology volume 3: The neuroscience of morality: Emotion, brain disorders, and development*. Cambridge, MA & London.: MIT Press, pp. 371–394.

Kant, I., [1785] 2002. *Groundwork for the metaphysics of morals*. New Haven: Yale University Press.

Kant, I., [1787] 1979. *Critique of pure reason*. Translated from German by Meiklejohn, J. M. D., London: J. M. Dent & Sons.

Kant, I., [1787] 1996. *Critique of practical reason*. Translated from German by Abbott, T. K., New York: Prometheus Books.

Kant, I., [1797] 2000. *The metaphysics of morals*. Translated from German by Gregor, M., Cambridge: Cambridge University Press.

Kant, I., 1902. *Kants Gesammelte Schriften* (AA) (Kant's Collected Writings, Akademie Edition). vols. 1–29. Berlin: DeGruyter/Reimer.

Kenny, A., 1963. *Action, emotion and will*. London, New York: Routledge & Kegan Paul.

Kitching, G., 2002. Introduction. In: Kitching, G., & Pleasants, N., eds., *Marx and Wittgenstein: Knowledge, morality and politics*. London: Routledge, pp. 1–19.

Knobe, J., & Nichols, S., 2008. *Experimental philosophy*. Oxford: Oxford University Press.

Krebs, D., 2011. *The origins of morality: An evolutionary account*. Oxford: Oxford University Press.

Kroeber, A. L., 1920. Totem and taboo: An ethnologic psychoanalysis. *American Anthropologist*, 22, pp. 48–55.

Kroeber, A. L., 1939. Totem and taboo in retrospect. *American Journal of Sociology*, 55, pp. 446–457.

Kronqvist, C., 2005. Emotions and our understanding of others and ourselves – 'The Most Beautiful Girl in the World'. In: Carlson, Å., ed., *Philosophical aspects on emotions*. Stockholm: Thales, pp. 295–304.

Kronqvist, C., 2008. *What we talk about when we talk about love*. Doctoral dissertation, Åbo: Åbo Akademi University.

Kronqvist, C., 2009. Our struggles with reality. In: Gustafsson, Y., Kronqvist, C., & McEachrane, M., eds., *Emotions and understanding: Wittgensteinian perspectives*. Basingstoke: Palgrave, pp. 202–220.

Kronqvist, C., 2014. The relativity of Westermarck's moral relativism. In: Shankland, D., ed., *Westermarck*. Occasional paper No. 44 of the Royal Anthropological Institute. Canon Pyon: Sean Kingston Publishing, pp. 124–144.

Kusch, M., 2014. Psychologism. In: Zalta, E. N., ed., *The Stanford encyclopedia of philosophy* (Spring 2014 edition). URL = <http://plato.stanford.edu/archives/spr2014/entries/psychologism/>.

Kymlicka, W., 2002. *Contemporary political philosophy: An introduction*. 2nd ed. Oxford: Oxford University Press.

Kymlicka, W., & Norman, W., 1994. Return of the citizen: A survey of recent work on citizenship theory. *Ethics*, 104(2), pp. 352–381.

Lagerborg, R., 1934. *Om sanningen i religionen och fromhetslivet*. Helsingfors: Söderström.

Lagerborg, R., 1951. *Edvard Westermarck och verken från hans verkstad under hans tolv sista år 1927–1939*. Helsingfors: Holger Schildts förlag.

Lagerborg, R., 1953. The essence of morals: Fifty years (1895–1945) of rivalry between French and English sociology. *Transactions of the Westermarck Society*, II, pp. 9–25.

Lagerspetz, O., 2012. Rape among scorpionflies, spouse abuse among the praying mantis, and other 'reproductive strategies' in the animal and human world. In: Forsberg, N., Burley, M., & Hämäläinen, N., eds., *Language, ethics and animal life: Wittgenstein and beyond*. London: Continuum, pp. 65–81.

Lagerspetz, O., 2013. Westermarck, Edward. In: McGee, R.J., & Warms, R. L., eds., *Theory in social and cultural anthropology: An encyclopedia*. Vol. II. Los Angeles: Sage, pp. 918–921.

Lagerspetz, O., & Suolinna, K., 2014. *Edward Westermarck: Intellectual networks, philosophy and social anthropology*. Helsinki: The Finnish Society of Arts and Letters, Commentationes scientiarum socialium, vol. 77.

Leach, E., 1981. Biology and social science: Wedding or rape? *Nature*, 291, pp. 267–268.

LeDoux, J., 1996. *The emotional brain*. New York: Simon and Schuster.

Lévi-Strauss, C., [1945] 1982. The work of Edward Westermarck. Translated by Alice & Timothy Stroup. In: Stroup, T., ed., *Edward Westermarck: Essays on his life and works*,

Acta Philosophica Fennica vol. 34, pp. 181–194. Originally published in *Revue de l'histoire des réligions* 129 (1945: 1 & 2–3, pp. 84–100).

Lévi-Strauss, C., 1969. *The elementary structures of kinship*. Boston: Beacon Press.

Lewontin, R. C., 1981. Sleight of hand. [Review of Lumsden & Wilson, 1981]. *The Sciences*, 21, pp. 23–26.

Lewontin, R. C., Rose, S., & Kamin, L., 1984. *Not in our genes*. New York: Pantheon Books.

Lieberman, D., Antfolk. J., & Santtila, Pl, submitted. Mommy's baby, daddy's maybe – how men assess paternity.

Lieberman, D., & Lobel, T., 2011. Kinship on the Kibbutz: Coresidence duration predicts altruism, personal sexual aversions and moral attitudes among communally reared peers. *Evolution and Human Behavior*, 33(1), pp. 26–34.

Lieberman, D., Tooby, J., & Cosmides, L., 2003. Does morality have a biological basis?: An empirical test of the factors governing moral sentiments relating to incest. *Proceedings of the Royal Society: Biological Sciences*, 270(1517), pp. 819–826.

Lieberman, D., Tooby, J., & Cosmides, L., 2007. The architecture of human kin detection. *Nature*, 445(7129), pp. 727–731.

Lieberman, D. & Antfolk, J. (2015). Human sexuality and inbreeding avoidance. In Buss, D. (ed.) *The handbook of Evolutionary Psychology*, John Wiley & Sons, New York, pp. 444–461.

Lipps, T., 1897. *Raumästhetik und geometrisch-optische Täuschungen*. Leipzig: JA Barth.

Lumsden, C. J., & Wilson, E. O., 1980. Gene-culture translation in the avoidance of sibling incest. *Proceedings of the National Academy of Science*, 77(10), pp. 6248–6250.

Lumsden, C. L., & Wilson, E. O., 1981. *Genes, mind and culture: The coevolutionary process*. Cambridge, MA: Harvard University Press.

Lumsden, C. L., & Wilson, E. O., 1983. *Promethean fire*. Cambridge, MA: Harvard University Press

Lyons, A. P., & Lyons, H. D., 2004. *Irregular connections: A history of anthropology and sexuality*. Lincoln: University of Nebraska Press.

MacIntyre, A., 1973. The essential contestability of some social concepts. *Ethics*, 84(1), pp. 1–9.

MacIntyre, A., 2007. *After virtue*. Notre Dame: University of Notre Dame Press.

Malinowski, B., 1912. The economic aspect of the Intichiuma ceremonies. In: Castrén, O., Hirn, Y., Lagerborg, R., & Wallensköld, A., eds., *Festskrift tillegnad Edvard Westermarck den 20 november 1912*. Helsingfors: J. Simelii arvingars boktryckeri, pp. 81–108.

Malinowski, B., 1913. [Review of Émile Durkheim,] Les formes élémentaires de la vie religieuse. *Folk-lore*, 24, pp. 525–531.

Malinowski, B., 1922a. *Argonauts of the Western Pacific: An account of native enterprise and adventure in the archipelagoes of Melanesian New Guinea: With a preface by Sir James George Frazer, F.B.A, F.R.S.* London: Routledge.

Malinowski, B., 1922b. Sexual life and marriage among primitive mankind. [Review of HHM I-III]. *Nature*, 109(2738), pp. 126–130.

Malinowski, B., 1923. Psychoanalysis and anthropology. [Letter to the Editor]. *Nature*, 112(2818), pp. 650–651.

Malinowski, B., 1924. Psychoanalysis and anthropology. *Psyche*, 4, pp. 293–332.

Malinowski, B., 1927a. Anthropology of the westernmost Orient [Review of RBM I-II]. *Nature*, 110(17 December 1927), pp. 867–868.

Malinowski, B., 1927b. *Sex and repression in savage society*. London: Routledge & Kegan Paul.

Malinowski, B., 1929. *The sexual life of savages in northwestern Melanesia: An ethnographic account of courtship, marriage and family life among the natives of the*

Trobriand Islands, British New Guinea. With a preface by Havelock Ellis. London: Routledge & Kegan Paul.

Malinowski, B., 1935. *Coral gardens and their magic. A study of the methods of tilling the soil and of agricultural rites in the Trobriand Islands.* Vol. 1: *The description of gardening.* London: George Allen & Unwin.

Malinowski, B., [1944] 1969. *A scientific theory of culture and other essays.* London: Oxford University Press.

Marett, R. R., 1920. Psycho-analysis and the savage. *Athenæum*, 30, pp. 205–206.

May, R. M., 1979. When to be incestuous. *Nature*, 279(5710), pp. 192–194.

Maynard Smith, J., & Warren, N., 1982. Models of cultural and genetic change. *Evolution*, 36, pp. 620–627.

McCabe, J., 1983. FBD marriage: Further support for the Westermarck hypothesis of the incest taboo. *American Anthropologist*, 85, pp. 50–69.

McCann, H., 2006. Resisting naturalism: The case of free will. In: Corradini, A., Galvan, S., & Lowe, E. J., eds., *Analytic philosophy without naturalism.* London and New York: Routledge, pp. 225–240.

McDowell, J., 1998. *Mind, value, and reality.* Cambridge, MA: Harvard University Press.

McKay, R. T., & Dennett, D. C., 2009. The evolution of misbelief. *Behavioral and Brain Sciences*, 32, pp. 493–561.

Meixner, U., 2006. Consciousness and freedom. In Corradini, A., Galvan, S., & Lowe, E. J., eds., *Analytic philosophy without naturalism.* Abingdon: Routledge, pp. 183–196.

Moll, A., 1908. *Das Sexualleben des Kindes.* Leipzig: F.C.W. Vogel.

Moody-Adams, M., 1997. *Fieldwork in familiar places: Morality, culture and philosophy.* Cambridge, MA: Harvard University Press.

Moore, G. E., 1903. *Principia ethica.* Cambridge: Cambridge University Press.

Moore, G. E., 1922. *Philosophical studies.* London: Routledge & Kegan Paul Ltd.

Morrisson, I. P. D., 2008. *Kant and the role of pleasure in moral acting.* Athens: Ohio University Press.

Murdoch, I., 1997a. Metaphysics and ethics. In: Conradi, P. J., ed., *Existentialists and mystics: Writings on philosophy and literature.* London: Chatto & Windus, pp. 59–75.

Murdoch, I., 1997b. The idea of perfection. In: Conradi, P. J., ed., *Existentialists and mystics: Writings on philosophy and literature.* London: Chatto & Windus, pp. 299–335.

Murdock, G. P., 1949. *Social Structure.* New York: Macmillan.

Nash, G. H., 2006. *The conservative intellectual movement in America since 1945.* Thirtieth-anniversary edition. Wilmington: ISI Books.

Nesse, R. M., 2000. How selfish genes shape moral passions. *Journal of Consciousness Studies*, 7, pp. 227–231.

Nichols, S., 2004. *Sentimental rules: On the natural foundations of moral judgement.* Oxford: Oxford University Press.

Nietzsche, F., [1887] 1981. Zur Genealogie der Moral. In: Nietzsche, F., ed., *Werke in zwei Bänden.* München: Carl Hanser Verlag, pp. 175–288.

Noë, A., 2010. *Out of our heads: Why you are not your head and other lessons from the biology of consciousness.* New York: Hill & Wang.

Norton, D. F., & Kuehn, M., 2006. The foundations of morality. In: Haakonssen, K., ed., *The Cambridge history of Eighteenth-century philosophy.* Vol. II. Cambridge: Cambridge University Press, pp. 939–986.

Nussbaum, M. C., 2001. *Upheavals of thought: The intelligence of emotions.* Cambridge: Cambridge University Press.

Nykänen, H., 2014. Conscience and collective pressure. *Philosophy, Psychiatry & Psychology*, 21, pp. 51–65.

Olsen, J. A., 2013. Neo-Darwinian conservatism in the United States. Doctoral dissertation. Oslo: University of Oslo.

Österman, T., 2014. *Rationality and cultural understanding*. Doctoral dissertation, Uppsala: Uppsala University.

Parker, H., & Parker, S., 1986. Father-daughter incest: An emerging perspective. *The American Journal of Orthopsychiatry*, 56(4), pp. 531–549.

Parsons, A., 1964. Is the Oedipus complex universal?: The Jones-Malinowski debate revisited and a south Italian 'nuclear complex'. In: Muensterberger, W., & Axelrad, S., eds., *The psychoanalytic study of society*. Vol. 3. New York: International Universities Press, pp. 278–328.

Paul, A., & Kuester, J., 2004. The impact of kinship on mating and reproduction. In Chapais, B., & Berman, C. M., eds., *Kinship and behavior in primates*, Oxford: Oxford University Press, pp. 271–290.

Pear, T. H., 1960. Some early relations between English ethnologists and psychologists. *Journal of the Royal Anthropological Institute*, 90, pp. 227–237.

Pinker, S., 2012. The false allure of group selection. *Edge* Website, available at: https://www.edge.org/conversation/steven_pinker-the-false-allure-of-group-selection [accessed 30 May 2016].

Pipatti, O., 2011. Edvard Westermarckin moraalisosiologia. *Sosiologia*, 48(3), pp. 238–252.

Pipping, K., 1984. Who reads Westermarck today? *British Journal of Sociology*, 35(3), pp. 315–332.

Pratt, M. L., 1986. Fieldwork in common places. In Clifford, J., & Marcus, G., eds., *Writing culture: The poetics and politics of ethnography*. Berkeley: University of California Press, pp. 27–50.

Price, H., 2004. Naturalism without representationalism. In: De Caro, M., & Macarthur, D., eds., *Naturalism in question*. Cambridge, MA & London, England: Harvard University Press, pp. 71–88.

Prinz, J., 2006. The emotional basis of moral judgements. *Philosophical Explorations*, 9(1), pp. 29–43.

Prinz, J. J., 2007a. Can moral obligations be empirically discovered? *Midwest Studies in Philosophy*, 31(1), pp. 271–291.

Prinz, J., 2007b. *The emotional construction of morals*. Oxford: Oxford University Press.

Prinz, J. J., 2011. Is empathy necessary for morality? In: Coplan, A., & Goldie, P., eds., *Empathy: Philosophical and psychological perspectives*. Oxford: Oxford University Press, pp. 211–229.

Pusey, A., & Wolf, A. P., 2005. Inbreeding avoidance in animals. In: Wolf, A. P., & Durham, W. H., eds., *Inbreeding, incest, and the incest taboo: The state of knowledge at the turn of the century*. Stanford, CA: Stanford University Press, pp. 202–205.

Pusey, A., & Wolf, M., 1996. Inbreeding avoidance in animals. *Trends in Ecology and Evolution*, 11(5), pp. 201–206.

Quine, W. V. O., [1953] 1980. Two dogmas of empiricism. In: Quine, W. V. O., ed., 1980. *From a logical point of view*. 2nd revised edition. Cambridge, MA: Harvard University Press, pp. 20–46.

Quine, W. V. O., 1969. Natural kinds. In: Quine, W. V. O., ed., *Ontological relativity and other essays*. New York: Columbia University Press, pp. 114–138.

Raphael, D. D., 2007. *The impartial spectator: Adam Smith's moral philosophy*. Oxford: Oxford University Press.

Reich, W., 1994. *Beyond psychology: Letters and journals 1934–1939*. Edited and with an introduction by Higgins, M. B., Translated by Derek, & Jordan, I., and Schmitz, P. New York: Farrar, Straus & Giroux.

Reichmayr, J., 1994. *Spurensuche in der Geschichte der Psychoanalyse*. Frankfurt am Main: Fischer Taschenbuch Verlag.

Reichmayr, J., 1995. *Einführung in die Ethnopsychoanalyse: Geschichte, Theorien und Methoden*. Frankfurt am Main: Fischer Taschenbuch Verlag.

Reik, T., [1928] 1931. *Ritual: Psycho-analytic studies*. Translated by Bryan, D. London: Hogarth Press.

Rivers, W. H. R., 1916. Sociology and psychology. *Sociological Review*, 9, pp. 1–13.

Robin, C., 2011. *The reactionary mind: Conservatism from Edmund Burke to Sarah Palin*. Oxford: Oxford University Press.

Róheim, G., 1917. Spiegelzauber. *Imago*, 5, pp. 63–120.

Róheim, G., 1918. Psychoanalysis és ethnologia. *Ethnographia*, 30, pp. 49–90, 206–245.

Róheim, G., 1919. *Spiegelzauber*. Wien: Internationaler Psychoanalytischer Verlag.

Róheim, G., 1932. Psychoanalysis of primitive cultural types. *International Journal of Psycho-Analysis*, 13(1), pp. 1–224.

Rorty, R., 2010. Naturalism and quietism. In: De Caro, M., & Macarthur, D., eds., *Naturalism and normativity*. New York: Columbia University Press, pp. 55–68.

Rose, H. J., 1934. [Review of Westermarck 1934b]. *Folk-Lore*, 45(4), pp. 357–359.

Rosen, L., 2002. *The culture of Islam: Changing aspects of contemporary Muslim life*. Chicago: The University of Chicago Press.

Rouse, S., 1998. Expedition and institution: A.C. Haddon and anthropology at Cambridge. In: Herle, A., & Rose, S., eds., *Cambridge and the Torres Strait: Centenary essays on the 1898 anthropological expedition*. Cambridge: Cambridge University Press, pp. 50–76.

Ruse, M., & Wilson, E. O., 1986. Moral philosophy as applied science. *Philosophy*, 61, pp. 173–192.

Ruth, H. S. and Reitz, K. R., 2003. *The challenge of crime: Rethinking our response*. Cambridge, MA: Harvard University Press.

Ryle, G., 1955. *The concept of mind*. London: Hutchinson & Co. Publishers Ltd.

Sabour, M., 1993. La baraka: capital et pouvoir symbolique. In: Bourquia, R., & Al Harras, M., eds., *Westermarck et la société marocaine*. Rabat: Université Mohammed V, Publications de la Faculté des Lettres et des Sciences Humaines – Rabat. Série Colloques et Séminaires no 27, pp. 125–130.

Salter, F., 2008. Westermarck's altruism. *Politics and the Life Sciences*, 27(2), pp. 28–46.

Sanderson, S. K., 2006. Reforming theoretical work in sociology: A detailed reply to my critics. [pdf] *Perspectives*, the newsletter of the ASA theory section. Available at: http://www.csun.edu/~egodard/asatheory/docs/theoreticalplural.pdf [accessed 26 May 2015].

Sanderson, S. K., 2007. *Evolutionism and its critics: Deconstructing and reconstructing an evolutionary interpretation of human society*. Boulder: Paradigm Publishers.

Schopenhauer, A., [1841] 1919. Preisschrift über die Grundlage der Moral. In: Frauenstädt, J., Hrsg. *Arthur Schopenhauer's sämmtliche Werke, 4. II. Die beiden Grundprobleme der Ethik*. Leipzig: Brockhaus, pp. 103–276.

Seemanová, E., 1971. A study of children of incestuous matings. *Human Heredity*, 21(2), pp. 108–128.

Segerstrale, U., 2000. *Defenders of the truth: The battle for science in the sociobiology debate and beyond*. Oxford: Oxford University Press.

Segerstrale, U., 2002. Neo-Darwinism. In Pagel, M., ed., *Oxford encyclopedia of evolution*. Oxford: Oxford University Press, pp. 107–110.

Self, R.O., 2012. *All in the family: The realignment of American democracy*. New York: Hill and Wang.

Seligman, B. Z., 1935. The incest taboo as a social regulation. *Sociological Review*, 27, pp. 75–93.

Seligman, C. G., 1924. Anthropology and psychology. A study of some points of contact. *Journal of the Royal Anthropological Institute*, 54, pp. 13–46.
Sesardic, N., 1998. From biological inhibitions to cultural prohibitions: How not to refute Edward Westermarck. *Biology and Philosophy*, 13, pp. 413–426.
Sesardic, N., 2005. From genes to incest taboos: The crucial step. In: Wolf, A. P., & Durham, W. H., eds., *Inbreeding, incest, and the incest taboo: The state of knowledge at the turn of the century*. Stanford, CA: Stanford University Press, pp. 109–120.
Shand, A. F., 1914. *The foundations of character: Being a study of the tendencies of the emotions and sentiments*. London: Macmillan & Co.
Shand, A. F., 1922. The relations of complex and sentiment. III. *British Journal of Psychology*, 13(2), pp. 123–129.
Shaftesbury, Third Earl Of (A. A. C.), [1711], 1999. *Characteristics of men, manners, opinions, times*. Ed. Lawrence E. Klein. Cambridge: Cambridge University Press.
Shankland, D., 2014a. Westermarck, anthropology and the Royal Anthropological Institute. In: Shankland, D., ed., *Westermarck*. Occasional paper No. 44 of the Royal Anthropological Institute. Canon Pyon: Sean Kingston Publishing, pp. 19–27.
Shankland, D., ed., 2014b. *Westermarck*. Occasional paper No. 44 of the Royal Anthropological Institute. Canon Pyon: Sean Kingston Publishing.
Shepher, J., 1971. Mate selection among second generation kibbutz adolescents and adults: Incest avoidance and negative imprinting. *Archives of Sexual Behavior*, 1(4), pp. 293–307.
Shepher, J., 1983. *Incest: A biosocial view*. New York: Academic Press.
Shostak, M., 1981. *Nisa: The life and words of a !Kung woman*. Cambridge, MA: Harvard University Press.
Smith, A., [1790] 1982. *The theory of moral sentiments*. Indianapolis: Liberty Fund.
Smith, J.M., 1982. *Evolution and the theory of games*. Cambridge: Cambridge University Press.
Smith, W. R., [1889] 1894. *Lectures on the religion of the Semites*. New [second] edition. London: A. & C. Block.
Sober, E., & Wilson, D. S., 2003. *Unto others: The evolution and psychology of unselfish behavior*. Cambridge, MA: Harvard University Press.
Solomon, R. C., 1980. Emotions and choice. In: Oksenberg Rorty, A., ed., *Explaining emotions*. Berkeley, LA & London: University of California Press, pp. 251–281.
Spencer, W. B., & Gillen, F. J., 1899. *The native tribes of central Australia*. London: Macmillan.
Sperber, D., 1985. *On anthropological knowledge: Three essays*. New York: Cambridge University Press.
Spiro, M. E., 1982. *Oedipus in the Trobriands*. Chicago: The University of Chicago Press.
Spiro, M. F., 1958. *Children of the kibbutz*. Cambridge: Harvard University Press.
Stevenson, C. S., 1937. The emotive meaning of ethical terms. *Mind*, 46(181), pp. 14–31.
Stocking, G. W., Jr., 1986. Anthropology and the science of the irrational: Malinowski's encounter with Freudian psychoanalysis. In: Stocking, G.W. Jr., ed., *Malinowski, Rivers, Benedict and others. Essays on culture and personality. History of anthropology*. Vol. 4. Madison, WI: The University of Wisconsin Press, pp. 13–49.
Stroup, T., 1981. In defense of Westermarck. *Journal of History of Philosophy*, 19(2), pp. 213–234.
Stroup, T., 1982a. Soft subjectivism. In: Stroup, T., ed., *Edward Westermarck: Essays on his life and works*. Helsinki: Acta Philosophica Fennica 34, pp. 99–121.
Stroup, T., 1982b. *Westermarck's ethics*. Åbo: Åbo Akademi University Press.
Stroup, T., 1984. Westermarck: A reappraisal. *Man*, New Series 19(4), pp. 575–592.

Sulloway, F., 1979. *Freud: Biologist of the mind: Beyond the psychoanalytic legend.* New York: Basic Books.

Sulloway, F., 1991. Reassessing Freud's case histories. *Isis*, 82, pp. 245–275.

Sulloway, F., 1992. *Freud: Biologist of the mind: Beyond the psychoanalytic legend.* revised edition. Cambridge, MA: Harvard University Press.

Suolinna, K., 1993. Rituals of passage in Islamic folk culture – A comparison between studies by Edward Westermarck and Hilma Granqvist. In: Bourquia, R., & Al Harras, M., eds., *Westermarck et la société marocaine*. Rabat: Université Mohammed V, Publications de la Faculté des Lettres et des Sciences Humaines – Rabat. Série Colloques et Séminaires no 27, pp. 43–51.

Suolinna, K., & Lahtinen, T., 2003. Holiness, holy places and the El-Baqqali family in northern Morocco. *Temenos*, 37–38, pp. 209–226.

Tal, I., & Lieberman, D., 2007. Kin detection and the development of sexual aversion: Toward an integration of theories on family sexual abuse. In Salmon, C. A., & Schackelford, T. K., eds., *Family relationships – An evolutionary perspective*. New York: Oxford University Press, pp. 205–229.

Talmon, Y., 1964. Mate selection in collective settlements. *American Sociological Review*, 29(4), pp. 491–508.

Taylor, G., 1975. Justifying the emotions. *Mind*, 84(335), pp. 390–402.

Taylor, J., 2011. Moral sentiment and the sources of moral identity. In: Bagnoli, C., ed., *Morality and the emotions*. Oxford: Oxford University Press, pp. 257–274.

Taylor, J., 2014. Moral sense and moral sentiment. In: Garrett, A., ed., *The Routledge companion to eighteenth century philosophy*. New York: Routledge, pp. 421–441.

Thompson, E., 2001. Empathy and consciousness. In: Thompson, E., ed., *Between ourselves: Second-person issues in the study of consciousness*. Thorverton: Imprint Academic, pp. 1–32.

Tiger, L., & Fox, R., 1971. *The imperial animal*. New York: Holt, Rinehart, Winston.

Titchener, E. B., 1909. *Lectures on the experimental psychology of thought processes*. New York: Macmillan.

Timosaari, N., 2010. Sukupuoli ja sivilisaatio Edvard Westermarckin ajattelussa. *Faravid*, 34, pp. 227–239.

Tinbergen, N., 1963. On the aims and methods of ethology. *Zeitschrift für Tierpsychologie*, 20, pp. 410–433.

Tooby, J., Cosmides, L., & Barrett, C., 2005. Resolving the debate on innate ideas. In: Carruthers, P., Laurence, S., & Stich, S., eds., *The innate mind.* Oxford: Oxford University Press, pp. 305–337.

Tooby, J., Cosmides, L., Sell, A., Lieberman, D., & Sznycer, D., 2008. Internal regulatory variables and the design of human motivation: A computational and evolutionary approach. In Elliot, A., ed., *Handbook of approach and avoidance motivation*. London: Taylor & Francis Group, pp. 251–271.

Trivers, R., 1972. Parental investment and sexual selection. In Krebs, J., & Davies, N., eds., *Sexual selection and the descent of man*. Boston: Blackwell, pp. 1871–1971.

Turco, L., 2003. Moral sense and the foundations of morals. In: Broadie, A., ed., *The Cambridge companion to the Scottish Enlightenment*. Cambridge: Cambridge University Press, pp. 136–156.

van den Berghe, P. L., 1980. Incest and exogamy: A sociobiological reconsideration. *Ethology and Sociobiology*, 1(2), pp. 151–162.

van den Berghe, P. L., 1983. Human inbreeding avoidance: Culture in nature. *Behavioral and Brain Sciences*, 6, pp.91–102.

van der Weele, C., 2011. Empathy's purity, sympathy's complexities: De Waal, Darwin and Adam Smith. *Biology and Philosophy*, 26, pp. 583–593.

von Wright, G. H., 1971. *Explanation and understanding*. New York: Cornell University Press.

von Wright, G. H., 1982. The origin and development of Westermarck's moral philosophy. In: Stroup, T., ed., *Edward Westermarck: Essays on his life and works*. Helsinki: Acta Philosophica Fennica 34, pp. 25–62.

Wallace, A. R., 1891. Letters to Edward Westermarck. Retrieved from http://people.wku.edu/charles.smith/wallace/S712.htm.

Wallace, E. R., 1983. *Freud and anthropology: A history and reappraisal*. New York: International Universities Press.

Warms, R. L., 2013. Tylor, Edward Burnett. In: McGee, R. J., & Warms, R. L., eds., *Theory in social and cultural anthropology: An encyclopedia*. London: SAGE, pp. 871–875.

Wikman, K. R. V., 1940. Letters from Edward B. Tylor and Alfred Russel Wallace to Edward Westermarck. *Acta academiae aboensis*, Humaniora, XIII(7), pp. 1–22.

Williams, B., 1985. *Ethics and the limits of philosophy*. London: Fontana Press.

Williams, G., 1996. *Adaption and natural selection*. Princeton: Princeton University Press.

Williams, L. M., & Finkelhor, D., 1995. Paternal caregiving and incest: Test of a biosocial model. *The American Journal of Orthopsychiatry*, 65(1), pp. 101–113.

Wilson, E. O., 1975. *Sociobiology: The new synthesis*. Cambridge, MA: Harvard University Press.

Wilson, E. O., 1978. *On human nature*. Cambridge, MA: Harvard University Press.

Wilson, E.O., 1998. *Consilience: The unity of knowledge*. New York: Alfred Knopf.

Wilson, E. O., 2012. *The social conquest of Earth*. New York: W. W. Norton.

Wilson, J. Q., 1993. *The moral sense*. New York: Free Press.

Wilson, J. Q., 2002. *The marriage problem: How our culture has weakened families*. New York: HarperCollins.

Winch, P., [1958] 1990. *The idea of a social science and its relation to philosophy*. London: Routledge & Kegan Paul.

Winch, P., 1972a. Moral integrity. In: Winch, P., ed., *Ethics and action*. London: Routledge & Kegan Paul, pp. 171–192.

Winch, P., 1972b. Understanding a primitive society. In: Winch, P., ed., *Ethics and action*. London: Routledge & Kegan Paul, pp. 8–49.

Winch, P., 1988. Who is my neighbour? In: Hertzberg, L., & Pietarinen, J., ed., *Perspectives on human conduct*. Leiden: E.J. Brill, pp. 97–109.

Wittgenstein, L., [1953] 2009. *Philosophical investigations*. Malden: Blackwell Publishing.

Wittgenstein, L., 1998. *Remarks on the philosophy of psychology*, Vol. II. Eds. von Wright, G. H., & Nyman, H., Translated from German by Luckhardt, C. G., & Aue, M. A. E., Oxford: Basil Blackwell.

Wittgenstein, L., 1999. *Philosophical occasions*. Edited by Klagge, J. C., & Nordman, A. Indianapolis: Hackett Publishing Company, Inc.

Wolf, A. P., 1966. Childhood association, sexual attraction, and the incest taboo: A Chinese case. *American Anthropologist*, 68, pp. 883–898.

Wolf, A. P., 1970. Childhood association and sexual attraction: A further test of the Westermarck Hypothesis. *American Anthropologist*, 72(3), pp. 503–515.

Wolf, A. P., 1993. Westermarck redivivus. *Annual Review of Anthropology*, 22, pp. 157–175.

Wolf, A. P., 1995. *Sexual attraction and childhood association*. Stanford: Stanford University Press.

Wolf, A. P. 2011. Westermarck and the Westermarck hypothesis. In: Shankland, D., ed., *Westermarck*. Occasional paper No. 44 of the Royal Anthropological Institute. Canon Pyon: Sean Kingston Publishing, pp. 96–103.

Wolf, A. P., & Durham, W. H., eds., 2005. *Inbreeding, incest, and the incest taboo: The state of knowledge at the turn of the century*. Stanford: Stanford University Press.

Wolf, A. P., & Huang, C. S., 1980. *Marriage and adoption in China, 1845–1945*. Stanford: Stanford University Press.

Young, M. W., 2004. *Malinowski: Odyssey of an anthropologist 1884–1920*. New Haven & London: Yale University Press.

Zahavi, A., 2000. Altruism: The unrecognized selfish traits. *Journal of Consciousness Studies*, 7, pp. 253–256.

Zubaida, S., 2010. Is there a Muslim society? Ernest Gellner's sociology of Islam. In: Zubaida, S., ed., *Beyond Islam: A new understanding of the Middle East*. London: I.B.Tauris, London, pp. 31–76.

Index